AF430916

Digital Electronics
A Comprehensive Lab Manual

Digital Electronics
A Comprehensive Lab Manual

Dr. Cherry Bhargava

Associate Professor and Head-VLSI Domain
School of Electronics and Electrical Engineering
Lovely Professional University
Phagwara, India.

 **BS Publications**

A unit of **BSP Books Pvt. Ltd.**
4-4-309/316, Giriraj Lane, Sultan Bazar,
Hyderabad - 500 095 - T.S.

Digital Electronics: A Comprehensive Lab Manual
by Dr. *Cherry Bhargava*

Published by:

 **BS Publications**
A unit of **BSP Books Pvt. Ltd.**
4-4-309/316, Giriraj Lane, Sultan Bazar,
Hyderabad - 500 095, T.S.
Phone : 040 - 23445688, 23445600
e-mail : info@bspbooks.net
Website : www.bspbooks.net

ISBN: 978-93-88305-95-2

PREFACE

Digital electronics is the foundation for digital computing and provides elemental knowledge on how circuits, systems and hardware communicate within the computer. The digital electronic is expanding on an accelerated rate and replacing the conventional analogue machines due to its remarkable features such as high accuracy, fast speed and flexibility. Ranging from miniature circuits to high end computers, the use of digital electronics is tremendous. It is therefore essential for engineers and students to understand the basics and implements the logic and design in consumer electronics and related fields.

The aim of this book is to teach you digital electronics at a practical level. By the end of this book, you should be able to combine circuit elements to create more complex circuits, and have an understanding of how it works. This handbook covers all the practicals related to combinational and sequential logic circuits as well as the significant questions for viva voce for examination preparation. As you progress through the book, you will learn about the building blocks of digital circuits and how they can be put together to build complex systems. It will improve the ease of conducting practicals with all required information available at one place along with detailed procedures of experiments. This book has been designed according to syllabi prescribed by AICTE in various universities.

I sincerely hope that this book proves to be useful and adequate to cater the needs of students. Any constructive feedback and suggestion from faculty members and students will be highly appreciated and will be incorporated in the next edition.

April 2019

Dr. Cherry Bhargava

CONTENTS

ABOUT THE AUTHOR

Dr. Cherry Bhargava obtained her PhD (ECE), IKGPTU, M.Tech (VLSI Design and CAD) from Thapar University. She received her B.Tech (Electronics and Instrumentation) from Kurukshetra University. She has qualified in GATE with All India Rank 428. She has more than 14 years of teaching and research experience. She is currently working as an associate professor and head, VLSI domain, School of Electrical and Electronics Engineering at Lovely Professional University, Punjab, India.

She has authored about 50 technical research papers in SCI, Scopus indexed quality journals and National / International conferences. She has four e-books to her credit. She has registered two copyrights and filed one patent. She is recipient of various National and International awards for being outstanding faculty in engineering and excellent researcher. She is an active reviewer and editorial member of various prominent SCI and Scopus indexed journals. She is a lifetime member of IET, IAENG, NSPE, IAOP, WASET and reliability research group. Her area of expertise includes reliability of electronic systems, digital electronics, VLSI design, artificial intelligence and related technologies.

Introduction of Digital IC and Equipments

OBJECTIVE

To familiarize with equipment and ICs used in digital electronics laboratory.

INTRODUCTION

Digital electronics is a branch of electronics which deals with arithmetic of digits, design of digital and logic circuits. To control and process various systems, digital circuits are mostly used. The term digit is derived from the digits. So, the systems, whose operation is based on digits, are categorized as digital system. Digital electronics involves the passage of electronic pulses along logic circuits.

DIGITAL INTEGRATED CIRCUITS (IC)

The word 'integration' signifies the collection of millions of electronic components and devices on a single chip. The integrated circuit is also abbreviated as an IC.

LEVEL OF INTEGRATION

Logic gates and memory devices are fabricated as integrated circuits (IC). With the advancement in technology, the level of integration enhances. The various active or passive components are interconnected within the chip, to form a digital circuit. The chip is mounted on the metal or plastic package, the connections are welded externally to form an IC.

Now a days, cost, size and speed are the main constraints of a designer and manufacturer. So, the number of components per chip is enhancing on the accelerating ratio. The table 1.1 shows the various level of integration, for an integrated circuit.

Table 1.1 IC Integration Level

Level of integration	Number of gates/chips
Small Scale Integration (SSI)	<12
Medium Scale Integration (MSI)	12-99
Large Scale Integration (LSI)	1000
Very Large Scale Integration (VLSI)	10k
Ultra Large Scale Integration (ULSI)	100k
Giga Scale Integration (GSI)	1 Meg

The logic gates IC is under SSI, combinational logic circuits are part of MSI and microprocessor is the part of LSI and VLSI level.

NOMENCLATURE OF IC

The manufacturer of integrated circuits, follow a standard numbering scheme, in which suffix, middle and prefix parts have some significance. For example, 7400 series IC and its derivatives are shown in fig.1.1.

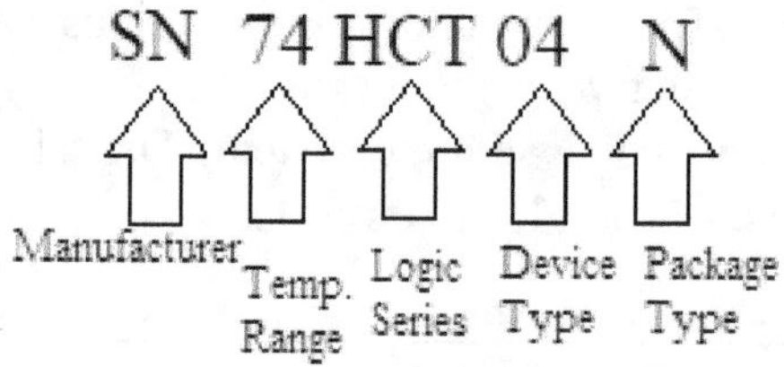

Fig. 1.1 IC number SN74HCT04N

Significance of IC number:

(a) *Manufacturer:* This code consists of two alphabets e.g. SN is used for Texas instruments, HEF for Mullard/Philips, DM for National semiconductors and S for Signetics.

(b) *Temperature range:* This code consists of two numeric. 74 signifies temperature range 0-70°C commercial, 54 military –55 to 125 °C.

(c) *Logic series:* This signifies sub family. 7400 series are most widely used in consumer applications.

(d) *Device type:* This indicates device type of function. For example, 04 indicates hex/inverter.

(e) *Package type:* It signifies the package type of IC for example, N is for plastic dual in line, W is for ceramic flat pack, D surface mounted plastic package.

FEW COMMONLY USED DIGITAL IC

The commonly used digital integrated circuits are listed as in Table 1.2.

Table 1.2 Digital Circuit IC Numbers

S. No	Digital Logic	Parameter	IC/Board Number
1		Quad 2-input AND logic gate	7408
2		Quad 2-input OR logic gate	7432
3		NOT logic gate/hex inverter	7404
4	Logic gates	Quad 2-input NAND logic gate	7400
5		Quad 2-input NOR logic gate	7402
6		Quad 2-input Exclusive OR logic gate	7486
7		Quad 2-input Exclusive NOR logic gate	74266 (TTL) 4077 (CMOS)
8		2:1 Multiplexer	74157
9	Multiplexer	4:1 Multiplexer	74153
10		8:1 Multiplexer	74151
11		16:1 Multiplexer	74150
12		1:2 Demultiplexer	74LVC1G19
13	Demultiplexer	1:4 Demultiplexer	74139
14		1:8 Demultiplexer	74138
15		1:16 Demultiplexer	74154
16		2: 4 Decoder	74155 (TTL)
17		3:8 Decoder	74137/74138
18	Decoder	4:16 Decoder	74154
19		BCD to decimal decoder	7441
20		BCD to seven segment decoders	7446/7447
21	Encoder	8:3 Priority Encoder	74148
22		10:4 Priority Encoder	74147
23	Digital	4-bit magnitude Comparator	7485
24	Comparator	8-bit magnitude Comparator	74682
25		SR Flip-flop	74279
26		JK Flip-flop	7470
27	Flip-flop	JK Master Slave Flip-flop	7471
28		D Flip-flop	7474/7479
29		T Flip-flop	7473 short J & K
30		8-bit Serial-In-Serial-Out register (SISO)	7491
31		8-bit Serial-In-Parallel-Out register (SIPO)	74164
32	Shift register	16-bit Parallel-in-Serial-Out register (PISO)	74674
33		4-bit Parallel-in-Parallel-Out register (PIPO)	7495
34		16-bit A/D converter	ADS5482 (TI)
35	ADC and DAC	16- bit D/A converter	DAC8728 (TI)
36		2-bit Full Adder	7482
37	Adder &	4-bit Full Adder	7483
38	Subtractor	4-bit Full Subtractor	74385

Table 1.2 *Contd...*

S. No	Digital Logic	Parameter	IC/Board Number
39	Counter	Up-down binary counter	74191
40		Up-down decade counter	74190
41		Modulo 10 counters	74416
42	Programmable Logic Devices (PLD)	Field Programmable Gate Array (FPGA)	SPARTAN 6 family, ARTIX 7 family
43		Complex Programmable Logic Device (CPLD)	ALTERA MAX 7000 series
44	Memories	16-bit RAM	7481/7484
45		64-bit RAM	7489
46		256-bit ROM	7488
47		512-bit ROM	74186 (open collector)
48		256-bit PROM with open collector output	74188
49		2048-bit PROM with open collector output	74470
50		2048-bit PROM with three state output	74471
51		1024-bit PROM with three state output	74287

Sometimes the chip manufacturer may denote the first pin by a small indented circle above the first pin of the chip. Pin 1 is often identified by a dot or a notch next to it on the chip package. Generally, an IC has four gates and each gate has two inputs and one output.

BREADBOARD

A breadboard is a device that is used to build connections of digital circuits. It is one of the most fundamental pieces when learning how to build circuits. A typical breadboard is shown as in fig. 1.2.

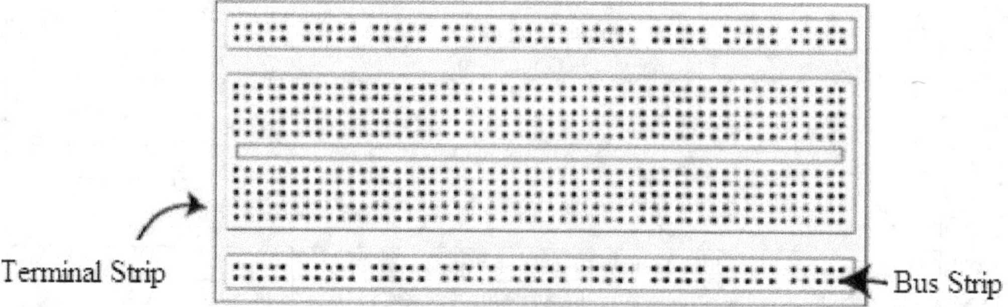

Fig. 1.2 Breadboard

Basically, the breadboard is consisted of two types of strips: terminal strip and bus strip. There is often a gap in between terminal and bus strip. Furthermore, each bus strip has two rows, each of the two rows of contacts are known as node,which are connected inside the breadboard. The bus strips are mostly used for power supply connections or in case when any node requires large number of connections.

The terminal strip is divided in two parts with a center. Each part has sixty rows and five columns. Each row of five contacts is known as node. The circuits can be built on the terminal strips by inserting the leads of circuit components into the contact receptacles and making connections with wire. It is a good practice to wire +5V and 0V power supply connections to separate bus strips.

DIGITAL IC TESTER

An Integrated Circuit tester (IC tester) is used to test Integrated Circuits (ICs). We can easily test any digital IC using this kind of an IC tester. It is necessary for the users to test the ICs before inserting them in respective project.

To test a particular digital IC, one needs to insert the IC into the IC socket and enter the IC number using the keyboard and then press the "ENTER" key. The IC number gets displayed in the 7-segment display unit.

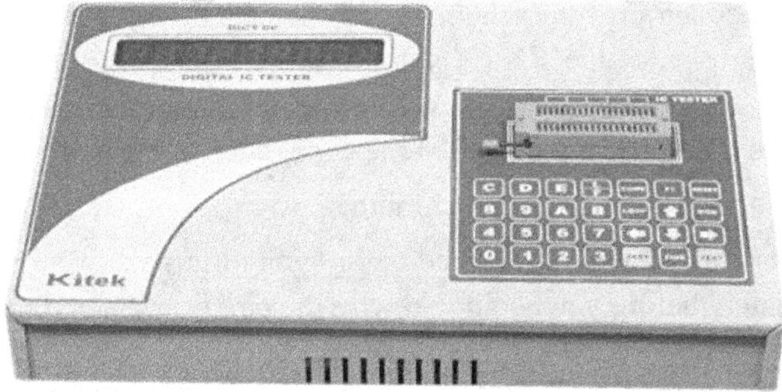

Fig. 1.3 Digital IC Tester

If the digital IC is faulty, it will display "FAIL" and if the digital IC is in working mode then it will display "PASS". It has potential free 40 pin universal ZIF socket. There is an audio alarm which indicates the user about faulty product.

POWER SUPPLY AND GROUND

The designer should connect the circuit with +5V for power supply and 0V for ground purpose. The voltage of power supply should not be exceeded beyond 5V to avoid any damage to ICs, during the conduct of experiments.

Incorrect connection of power to the ICs could result in them exploding or becoming extremely hot, with the possible serious injury occurring to the people working on the experiment. Ensure that the power supply polarity and all components and connections are correct before switching on power.

CIRCUIT DESIGNING PROCESS

The students should follow the following steps carefully, while designing the digital circuits in the laboratory.

1. Turn the power (Trainer Kit) off before you build anything.
2. Make sure the power is off before you build anything!
3. Connect the +5V and ground (GND) leads of the power supply to the power and ground bus strips on your breadboard.
4. Connect the components and ICs on the breadboard. Point all the ICs in the same direction with pin 1 at the upper-left corner.
5. Connect +5V and GND pins of each IC to the power and ground bus strips on the breadboard.
6. Select a connection on your schematic and place a piece of hook-up wire between corresponding pins of the ICs on your breadboard. It is better to make the short connections before the longer ones. Mark each connection on your schematic as you go, so as not to try to make the same connection again at a later stage.
7. Check the connections, before you turn the power on.
8. If an error is made and is not spotted before you turn the power on. Turn the power off immediately before you begin to rewire the circuit.

POST EXPERIMENT PROCESS

1. At the end of the laboratory session, remove the wires, ICs and all equipment and return them to the instructor.
2. Tidy the area that you were working in and leave it in the same condition as it was before you started.
3. In case, any equipment of IC has damaged during the experiment, do not put the faulty pieces in the equipment box. Convey the instructor about the same.

COMMON ERRORS

1. Not connecting the ground and/or power pins for all chips.
2. Not turning on the power supply before checking the operation of the circuit.

3. Loose connections.
4. Plugging wires into the wrong holes.
5. Driving a single gate input with the outputs of two or more gates.
6. Modifying the circuit with the power on.
7. Use of faulty equipment and ICs.
8. Use of wrong IC or component.

Study and Verification of Basic Logic Gates

OBJECTIVE

To study and verify the basic logic gates (AND, OR, NOT) using respective integrated circuits.

APPARATUS

Breadboard, Resistor 220 Ω, Power supply (+5V), LEDs, Connecting wires.

ICS REQUIRED

7408, 7432, 7404

INTRODUCTION

The logic gate is the building block of the digital system. One of more binary signals are inserted to the logic gate as an input and it produces a single output. Using logic gates, various logic circuits can be designed using transistors and resistors etc. The signals that are applied on input and output terminal of a logic gate can be at logic level HIGH or logic level LOW. The function of logic gate is expressed in terms of truth table. The truth table has all the possible combination of input variables and output variable.

The number of possible combinations 'N' of binary input to a logic gate is:

$$N = 2^n$$

where, n = number of bits

For 2-bit binary number, the possible combinations are $2^2 = 4$

For 3-bit binary number, the possible combinations are $2^3 = 8$

Each variable can be at either 0 or 1 i.e. logic level LOW or logic level HIGH

Logic gates are and electronic circuit elements which perform Boolean algebraic functions. There are total seven logic gates i.e. AND, NAND, OR, NOR, NOT, EX-OR, EX-NOR, out of which three are basic logic gates: AND, OR, NOT and two are universal logic gates: NAND and NOR. These gates are called universal gates because any logic gate can be derived from these gates.

BASIC LOGIC GATES

There are three logic gates: AND, OR & NOT, which are known as basic logic gates. The description and IC of these gates are described as below:

AND LOGIC GATE

The AND logic gate is a logic circuit whose output is 1 i.e. logic level HIGH when all the input variables are at level 1 i.e. at logic level HIGH.

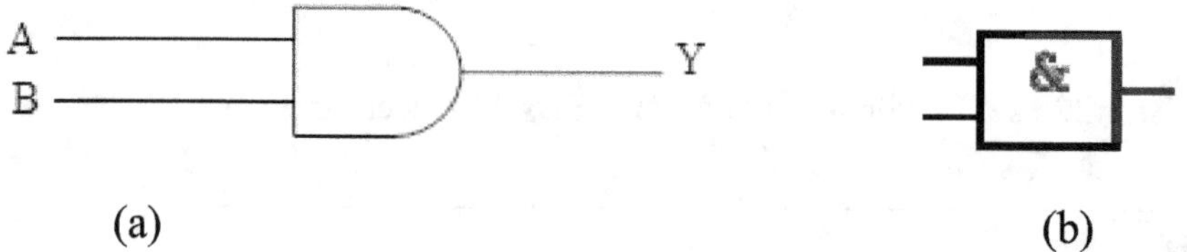

Fig 2.1 Symbol of AND Gate (a) MIL/ANSI Symbol (b) IEC symbol

The symbol of AND Gate is shown as in fig. 2.1. (a) and (b). In the AND function, two or more inputs are applied at input terminal and output is received at receiver terminal. The output is HIGH (Logic 1) only when all the inputs are high. The same is depicted in truth table 2.1.

The expression for AND gate is shown as:

$$Y = A \cdot B$$

The table which shows the correlation in between inputs and outputs, is known as truth table. The truth table of AND gate is shown in table 2.1.

Table 2.1 Truth Table of AND Gate

INPUT		OUTPUT
A	B	Y
0	0	0
0	1	0
1	0	0
1	1	1

When any of the input is LOW level i.e. 0, then output is also LOW level i.e. level 0. The output becomes HIGH only in the case when both the inputs are at HIGH level i.e. level 1.

OR LOGIC GATE

The OR logic gate is a type of logic gate, where output is HIGH i.e. Logic 1 when at least one input is HIGH i.e. Logic 1.

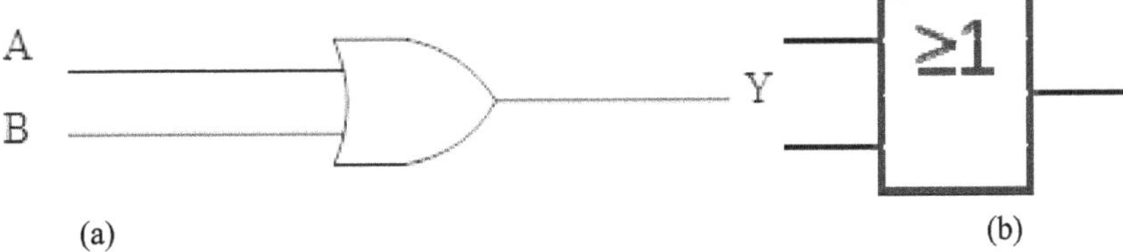

Fig 2.2 Symbol of OR Gate (a) MIL/ANSI Symbol (b) IEC symbol

The symbol of OR Gate is shown as in fig. 2.2. (a) and (b). In the OR function, two or more inputs are applied at input terminal and output is received at receiver terminal. The output is HIGH (Logic 1) only when at least one input is high. The same is depicted in truth table 2.2.

The expression for OR gate is shown as:

$$Y = A + B$$

The table which shows the correlation in between inputs and outputs, is known as truth table. The truth table of OR gate is shown in table 2.2.

Table 2.2 Truth Table of OR Gate

INPUT		OUTPUT
A	B	Y
0	0	0
0	1	1
1	0	1
1	1	1

When any of the input is HIGH level i.e. 1, then output is also HIGH level i.e. level 1. The output becomes LOW only in the case when both the inputs are at LOW level i.e. level 0.

NOT LOGIC GATE

A logic gate is said to be act as NOT gate or inverter when it changes the applied input logic level to opposite logic level. It is also called an inverting buffer.

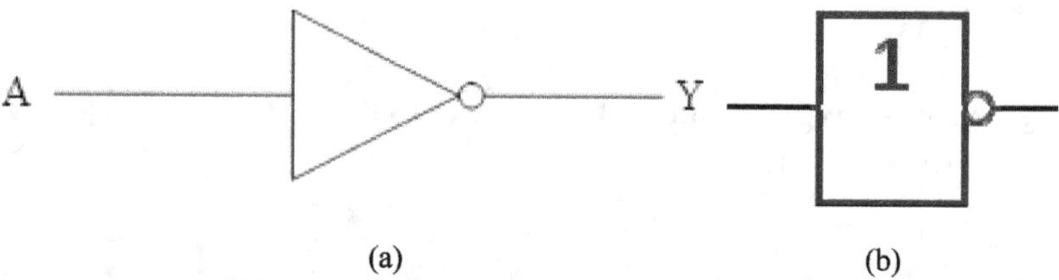

Fig 2.3 Symbol of NOT Gate (a) MIL/ANSI Symbol (b) IEC symbol

The symbol of NOT Gate is shown as in fig. 2.3. (a) and (b).

The NOT gate performs inversion or complement operation, where logic level of input variable is changed to opposite level. For example, if 0 is applied at input terminal, the output will provide value 1 and if 1 is applied at input terminal, the output will provide value 0. The same is depicted in truth table.

The expression for NOT gate is shown as:

$$Y = \bar{A}$$

The table which shows the correlation in between inputs and outputs, is known as truth table. The truth table of NOT gate is shown in table 2.3.

Table 2.3 Truth Table of NOT Gate

Input	Output
A	Y
0	1
1	0

When input is at HIGH level i.e. 1, then output will be at LOW level i.e. 0 and when input is at LOW level i.e. 0, then the output will be at HIGH level i.e. 1

PIN DIAGRAM OF AND, OR AND NOT GATE IC

The ICs used for AND, OR & NOT gate are 7408, 7432 and 7404 respectively. The pin diagrams of all the used ICs are depicted as below in fig. 2.4 to 2.6.

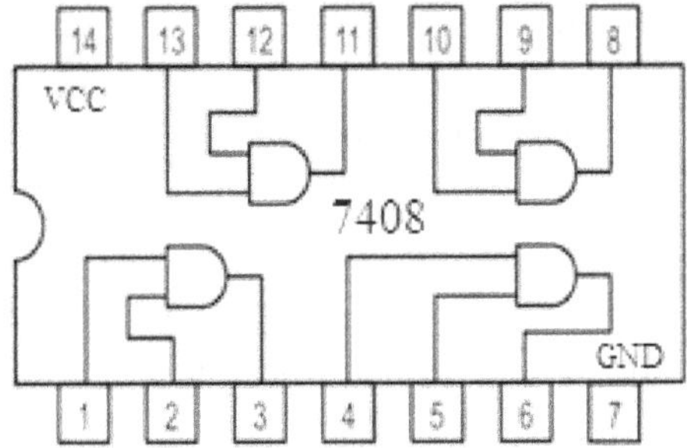

Fig. 2.4 Pin diagram of IC 7408 (AND gate)

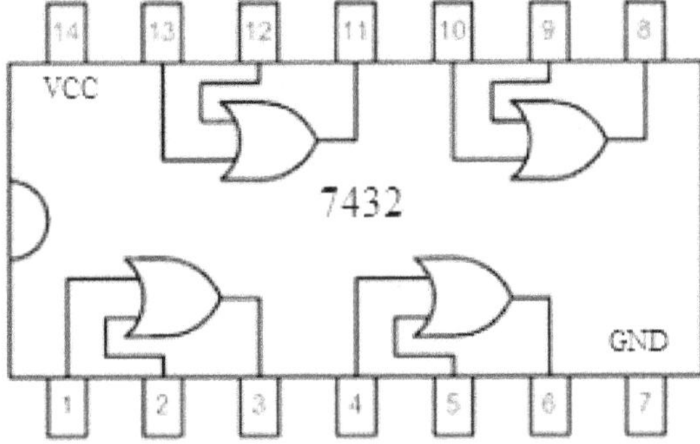

Fig. 2.5 Pin diagram of IC 7432 (OR gate)

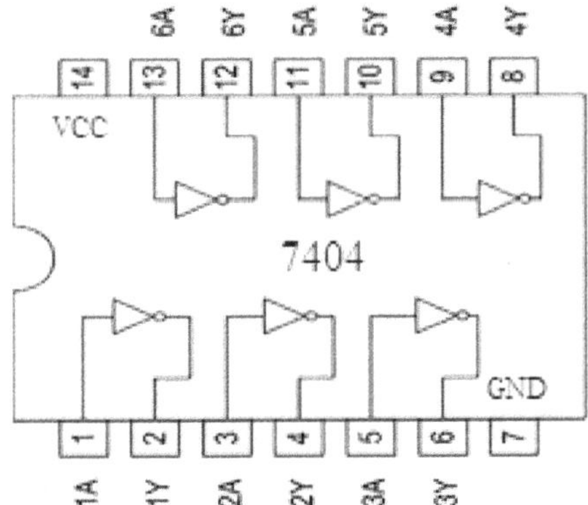

Fig. 2.6 Pin diagram of IC 7404 (NOT gate)

EXPERIMENTAL PROCESS

1. Check all the ICs using IC tester. After inserting IC carefully in ZIP socket of the tester, the tester should show "PASS".

2. To verify the truth table of AND, OR & NOT gate, insert the respective IC on the breadboard.

3. Connect the 5V (positive terminal) of power supply to pin 14 and 0V (negative terminal) to pin 7 of the IC.

4. To verify the inputs and outputs connection, connect 5V for logic 1/HIGH or 0V for logic 0/LOW.

5. Connect the output of logic gate to LED through a resistor of 220Ω.

6. Connect the inputs to various combinations, as described in truth table and observe the output.

7. For example, to verify the third condition of AND logic gate, the IC 7408 is taken. the pin 1 is connected to positive terminal of power supply i.e. 5V, pin 2 is connected to negative power supply i.e. 0V and pin 3 (output) is connected to positive terminal of LED through 220 Ω resistor and negative terminal of LED to 0V. As soon as power is switched ON, the LED will not glow, which satisfies the condition as described in truth table. Similarly, all other conditions can be verified by connecting the inputs accordingly.

8. The above procedure is repeated for all ICs and truth table of all logic gates are verified.

RESULT

The truth table of all the basic gates AND, OR & NOT is verified using respective ICs.

PRECAUTIONS

1. Handle the ICs carefully.
2. Check the ICs before starting of the experiment.
3. The connections should be neat and tight.
4. The LED has one leg long and other leg short. The long leg depicts positive end and short end depicts negative end.
5. Do not press the IC on breadboard until pins are aligned with pours properly.
6. Avoid short circuit and heating of ICs.

VIVA VOCE QUESTIONS

1. What is the significance of GND and Vcc in an IC?
2. What is the need of integration in modern electronics. Name any one of the devices with latest level of integration?

3. What is the significance of '74' in any IC?
4. If there are four inputs, make the truth table for (a) AND (b) OR logic gate?
5. Why AND, OR & NOT are called basic logic gates?
6. What are the applications of logic gates?
7. What is the 'logic' behind logic gate? Why it is called 'gate'?

Study and Verification of Universal Logic Gates

OBJECTIVE

To study and verify the Universal logic gates (NAND, NOR) using respective integrated circuits.

APPARATUS

Breadboard, Resistor 220 Ω, Power supply (+5V), LEDs, Connecting wires.

ICS REQUIRED

7400, 7402

INTRODUCTION

The logic gate is the building block of the digital system. One of more binary signals are inserted to the logic gate as an input and it produces a single output. Using logic gates, various logic circuits can be designed using transistors and resistors etc. The signals that are applied on input and output terminal of a logic gate can be at logic level HIGH or logic level LOW. The function of logic gate is expressed in terms of truth table. The truth table has all the possible combination of input variables and output variable.

The Logic gates are and electronic circuit elements which perform Boolean algebraic functions. There are total seven logic gates i.e. AND, NAND, OR, NOR, NOT, EX-OR, EX-NOR, out of which three are basic logic gates: AND, OR, NOT and two are universal logic gates: NAND and NOR. These gates are called universal gates because any logic gate can be derived from these gates.

UNIVERSAL LOGIC GATES

There are two universal logic gates: NAND & NOR, which are known as universal logic gates. These gates are called universal logic gates because any other logic gate can be derived with the help of these (NAND, NOR) gates. The description and IC of these gates are described as below:

NAND LOGIC GATE

A NAND gate is a logic gate whose output is LOW when all inputs are HIGH

From combination of basic gates AND, OR & NOT, more logic functions are derived. NAND gate is one of the advanced versions of basic gate which is derived from AND & NOT gate. This gate is called universal gate because any basic gate can be driven from this gate.

As NAND gate is combination of AND & NOT gate, it implies an AND function with complemented output. The symbol and its equivalence are shown as in fig. 3.1.

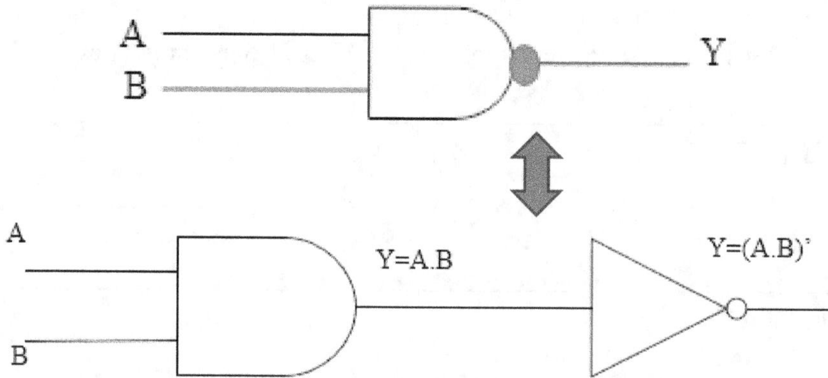

Fig. 3.1 NAND gate (a) symbol (b) equivalent

In the NAND gate logic function, two or more inputs are applied at input terminal and output is received at receiver terminal. The output is HIGH (Logic 1) when any of input variable is low. The same is depicted in truth table 3.1.

The expression for NAND gate is shown as:

$$Y = \overline{A.B}$$

The table which shows the correlation in between inputs and outputs, is known as truth table. The truth table of NAND gate is shown in table 3.1.

Table 3.1 Truth Table of NAND Gate

INPUT		OUTPUT
A	B	Y
0	0	1
0	1	1
1	0	1
1	1	0

When any of the input is LOW level i.e. level 0, then output will be at HIGH level i.e. level 1. The output becomes LOW only in the case when both the inputs are at HIGH level i.e. level 1.

NOR LOGIC GATE

A NOR gate is a logic gate where output is HIGH only when both inputs are LOW.

From combination of basic gates AND, OR & NOT, more logic functions are derived. NOR gate is one of the advanced versions of basic gate which is derived from OR and NOT gate. This gate is called universal gate because any basic gate can be driven from this gate.

As NOR gate is combination of OR and NOT gate, it implies an OR function with complemented output. The symbol and its equivalence are shown as in fig. 3.2.

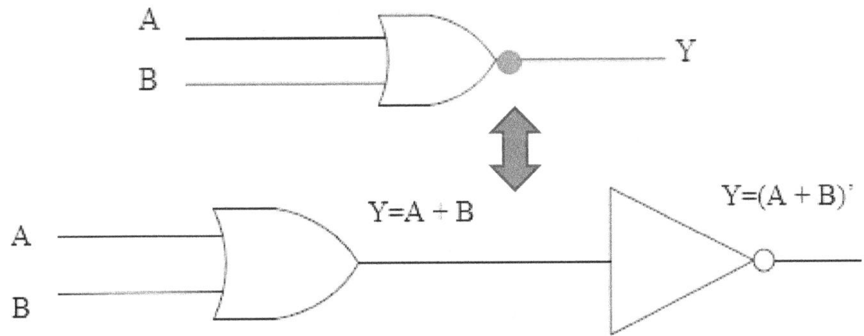

Fig. 3.2 NOR gate (a) symbol (b) equivalent

In the NOR gate logic function, two or more inputs are applied at input terminal and output is received at receiver terminal. The output is HIGH (Logic 1) when all input variables are low. The same is depicted in truth table 3.2. The expression for NOR gate is shown as:

$$Y = \overline{A + B}$$

The table which shows the correlation in between inputs and outputs, is known as truth table. The truth table of NOR gate is shown in table 3.2.

Table 3.2 Truth Table of NOR gate

INPUT		OUTPUT
A	**B**	**Y**
0	0	1
0	1	0
1	0	0
1	1	0

When any of the input is HIGH level i.e. level 1, then output will be at LOW level i.e. level 0. The output becomes LOW only in the case when both the inputs are at LOW level i.e. level 0.

PIN DIAGRAM OF NAND AND NOR GATE IC

The ICs used for NAND & NOR gate are 7400 and 7402 respectively. The pin diagrams of all the used ICs are depicted as below in fig. 3.3 and fig. 3.4.

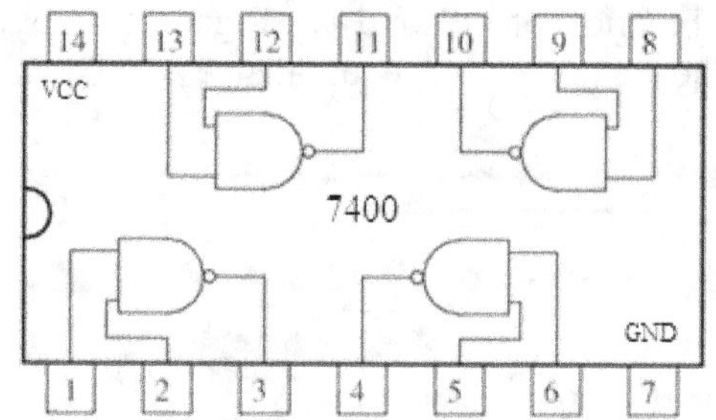

Fig. 3.3 Pin diagram of IC 7400 (NAND gate)

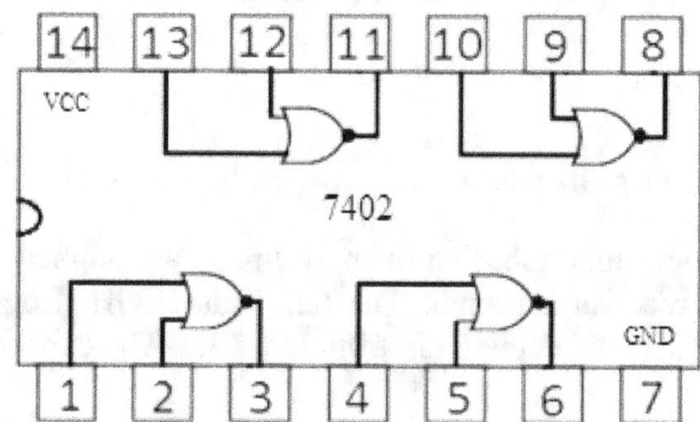

Fig. 3.4 Pin diagram of IC 7402 (NOR gate)

EXPERIMENTAL PROCESS

1. Check all the ICs using IC tester. After inserting IC carefully in ZIP socket of the tester, the tester should show "PASS".
2. To verify the truth table of AND, OR & NOT gate, insert the respective IC on the breadboard.
3. Connect the 5V (positive terminal) of power supply to pin 14 and 0V (negative terminal) to pin 7 of the IC.
4. To verify the inputs and outputs connection, connect 5V for logic 1/HIGH or 0V for logic 0/LOW.
5. Connect the output of logic gate to LED through a resistor of 220Ω.
6. Connect the inputs to various combinations, as described in truth table and observe the output.
7. For example, to verify the third condition of NAND logic gate, the IC 7400 is taken. the pin 1 is connected to positive terminal of power supply i.e. 5V, pin 2 is connected to negative power supply i.e. 0V and pin 3 (output) is connected to positive terminal of LED through 220 Ω resistor and negative terminal of LED to 0V. As soon as power is switched ON, the LED will glow, which satisfies the condition as described in truth table. Similarly, all other conditions can be verified by connecting the inputs accordingly.
8. The above procedure is repeated for all ICs and truth table of all logic gates are verified.

RESULT

The truth table of all the basic gates NAND & NOR is verified using respective ICs.

PRECAUTIONS

1. Handle the ICs carefully.
2. Check the ICs before starting of the experiment.
3. The connections should be neat and tight.
4. The LED has one leg long and other leg short. The long leg depicts positive end and short end depicts negative end.
5. Do not press the IC on breadboard until pins are aligned with pours properly.
6. Avoid short circuit and heating of ICs.

VIVA VOCE QUESTIONS

1. Why NAND and NOR are called Universal logic gates?
2. Design AND, OR & NOT using NAND logic gate?
3. Design AND, OR & NOT using NOR logic gate?
4. Elaborate the operation of NOR gate as negative AND gate?
5. Elaborate the operation of NAND gate as negative OR gate?
6. Specify the applications of NAND & NOR logic gates?
7. Give the expressions of Exclusive OR and Exclusive NOR gate?

Verification of De-Morgan's Theorem

OBJECTIVE

(A)   To study and verify the De-Morgan's theorem using logic gates.

(B)   Simplify Boolean expression using De-Morgan's theorem and verify it.

APPARATUS

Breadboard, Power supply (+5V), LEDs, Connecting wires.

ICS REQUIRED

7400, 7402, 7404, 7408, 7432

INTRODUCTION

Boolean algebra was first introduced by George Boole in 1847. In contrast with elementary algebra, the value of input variables in Boolean algebra is truth values i.e. true or false which are represented in digital system as 0 and 1. Boolean algebra is mathematical tool to analyse digital circuits.

The main motive of logic design is to minimize and simplify the logic so that minimum number of gates or wires are used, which will cause effect the cost, power dissipation and speed of that logic circuit. Using Boolean algebraic rules, the equations can be simplified and logic complexity can be further reduced.

DE-MORGAN'S THEOREM

A mathematician named De-Morgan's developed a pair of important rules regarding group complementation in Boolean algebra. It is based on the rule "Break the line and change the sign". De-Morgan's has suggested two theorems which are extremely useful in Boolean Algebra. The two theorems are discussed below. If there are two variables A and B, then as per De-Morgan's theorems:

Theorem 1: The complement of the sum of two or more variables is equal to the product of the compliment of the variables (fig. 4.1).

$$\overline{A + B} = \bar{A}.\bar{B}$$

Fig. 4.1 Circuit implementation of Theorem 1

Using perfect induction law, we can verify De-Morgan's theorem, here the values of AB are taken as 00, 01, 10 and 11 (Table 4.1).

Table 4.1 Verification of Theorem 1

A	B	$A + B$	$\overline{A + B}$	$\bar{A}$	$\bar{B}$	$\bar{A}.\bar{B}$
0	0	0	1	1	1	1
0	1	1	0	1	0	0
1	0	1	0	0	1	0
1	1	1	0	0	0	0

Hence Proved $\overline{A + B} = \bar{A}.\bar{B}$

Theorem 2: The complement of the product of two or more variables is equal to the sum of the compliment of the variables (fig. 4.2).

$$\overline{AB} = \bar{A} + \bar{B}$$

Fig. 4.2 Circuit implementation of Theorem 2

Using perfect induction law, we can verify De-Morgan's theorem, here the values of AB are taken as 00, 01, 10 and 11 (Table 4.2).

Table 4.2 Verification of Theorem 2

A	B	AB	$\overline{AB}$	$\bar{A}$	$\bar{B}$	$\bar{A} + \bar{B}$
0	0	0	1	1	1	1
0	1	0	1	1	0	1
1	0	0	1	0	1	1
1	1	1	0	0	0	0

Hence Proved $\overline{AB} = \bar{A} + \bar{B}$

So, De-Morgan's theorem is verified.

BOOLEAN EXPRESSION REDUCTION USING DE-MORGAN'S THEOREM

Take a Boolean expression $\overline{A + \overline{BC}}$. This will be reduced using De-Morgan's theorem.

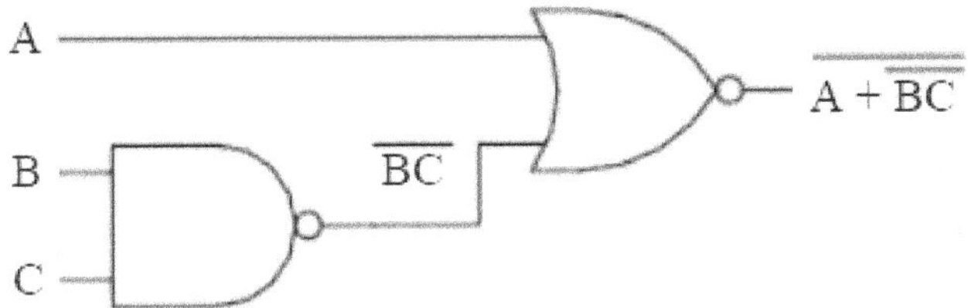

Fig. 4.3 Boolean expression

Using De-Morgan's theorem:

$$\overline{A + \overline{BC}} = \overline{A} + \overline{\overline{B} + \overline{C}} = \overline{A}.B.C$$

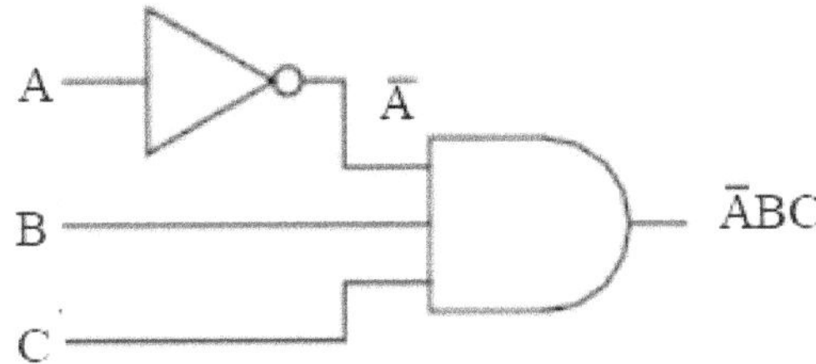

Fig. 4.4 Reduced Boolean expression

Boolean expression can be further verified using truth table as described in truth table 4.3.

Table 4.3 Boolean expression verification using De-Morgan's theorem

A	B	C	$\overline{A + \overline{BC}}$	$\overline{A}.B.C$
0	0	0		
0	0	1		
0	1	0		
0	1	1		
1	0	0		
1	0	1		
1	1	0		
1	1	1		

By using this method, De-Morgan's theorem is verified.

EXPERIMENTAL PROCEDURE

1. Check the functionality of ICs using IC tester.
2. Make the connections as per fig. 4.1 and 4.2. using various digital IC.
3. Verify the De-Morgan's theorem.
4. Connect +5V for power supply and 0V for ground purpose.
5. Verify the truth table 4.1 and 4.2.
6. Any Boolean expression can be verified using Boolean algebra.
7. The Boolean expression can be verified using truth table.

RESULT

1. De-Morgan's theorem is verified using basic logic gates.
2. Boolean expression is simplified using De-Morgan's theorem and verified using basic gates.

PRECAUTIONS

1. Handle the ICs carefully.
2. Check the ICs before starting of the experiment.
3. The connections should be neat and tight.
4. Do not press the IC on breadboard until pins are aligned with pours properly.
5. Avoid short circuit and heating of ICs.

VIVA VOCE QUESTIONS

1. What is the significance of Boolean operators?
2. How truth table is helpful in verifying the expression and logic?
3. What do you understand by minimised expression?
4. Simplify the Boolean expression $F = (A + B)(A + \bar{B}) + \overline{(A\bar{B})} + \bar{A}$ using De-Morgan's theorem?
5. What do you understand by term 'duality' in Boolean algebra?
6. Specify the Consensus theorem.
7. What do you mean by SOP and POS?

Binary Adder

OBJECTIVE

(A) To realize the Half Adder circuits using basic logic gates and to verify their truth tables.

(B) To realize the Full Adder circuits using basic logic gates and to verify their truth tables.

APPARATUS

Breadboard, Power supply (+5V), LEDs, Resistor 220Ω, Connecting wires.

ICS REQUIRED

7486, 7408, 7432

INTRODUCTION

The adder is a digital logic circuits which implements addition of numbers. In combinational logic circuits, an adder performs the addition of two binary bits. Depending on the number of bits, the adders are further classified in two types as half adder and full adder.

HALF ADDER

The half adder performs the operation of addition on two binary bits, augend and addend. It produces output in form of sum and carry (fig. 5.1).

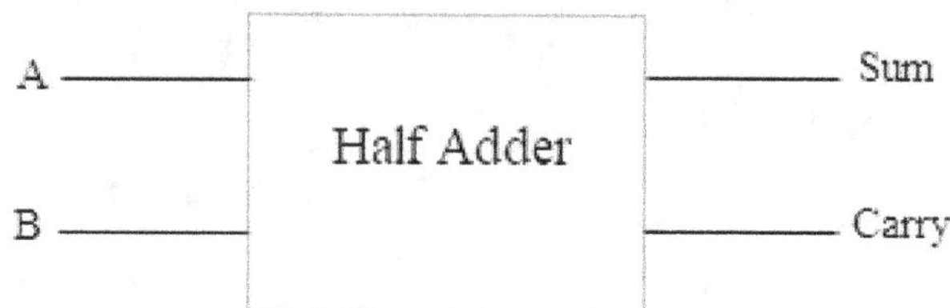

Fig. 5.1 Block diagram of Half Adder

The number of inputs and outputs are identified and labelled it. There are two inputs A and B producing two outputs Sum and carry. As the number of inputs are two, it suggests four possible combinations. The half adder performs binary addition.

Table 5.1 Truth table of Half adder

Inputs		Outputs	
A	B	Sum	Carry
0	0	0	0
0	1	1	0
1	0	1	0
1	1	0	1

A minimized Boolean function can be obtained using Karnaugh map. The two-variable K-map is used to explore the Boolean function.

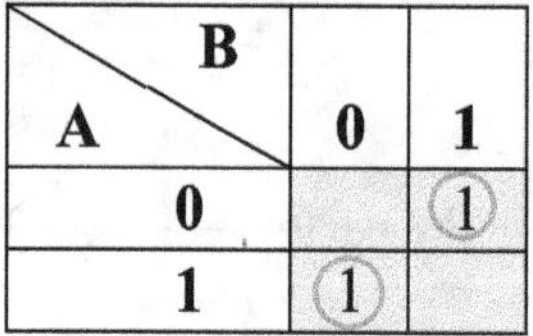 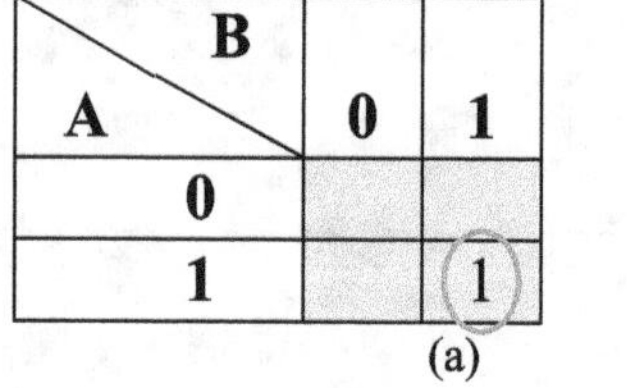

(a)   (b)

Fig. 5.2 K-map for (a) Sum and (b) Carry

Using K-map, as in fig. 5.2., we get,

$$\text{Sum} = A\bar{B} + \bar{A}B \tag{5.1}$$

$$\text{Carry} = AB \tag{5.2}$$

The logical representation of Sum and carry using combination of EX-OR and AND gate, is shown as in fig. 5.3.

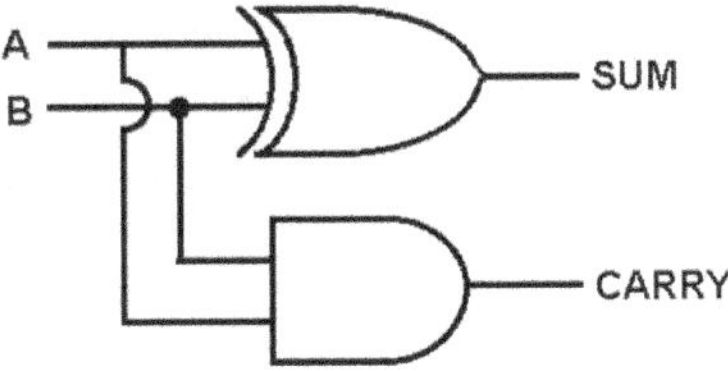

Fig. 5.3 Half Adder using logic gates

FULL ADDER

In order to accommodate the carry, generated by previous or low order bits, the full adder performs the addition of three binary bits. It consists of three inputs A, B, C_{in} and two outputs sum, carry (fig. 5.4)

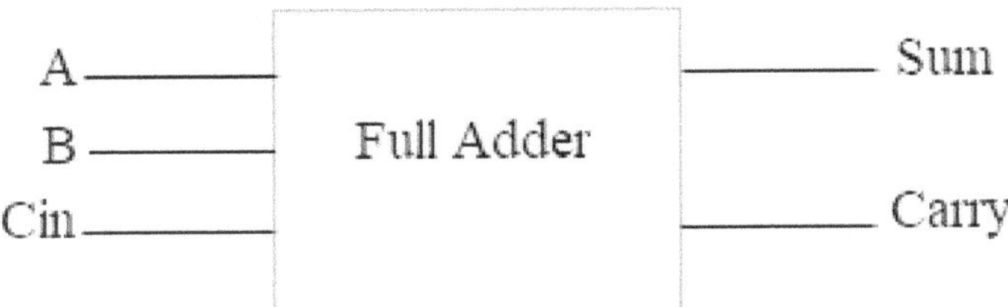

Fig. 5.4 Block diagram of Full Adder

The A and B are the significant binary bits and C_{in} is the carry from low order bit. As the number of inputs are three, so the possible combinations are eight. The truth table is formulated which shows the output sum and carry with respect to the all possible combinations of inputs. The table 5.2. shows the truth table of full adder.

Table 5.2 Truth table of full adder

Inputs			Outputs	
A	B	Cin	Sum	Carry
0	0	0	0	0
0	0	1	1	0
0	1	0	1	0
0	1	1	0	1
1	0	0	1	0
1	0	1	0	1
1	1	0	0	1
1	1	1	1	1

A minimized Boolean function can be obtained using Karnaugh map. The three-variable K-map is used to explore the Boolean function.

A \ BCin	00	01	11	10
0		(1)		(1)
1	(1)		(1)	

Fig. 5.5 K-map for sum

$$\text{Sum} = AB\overline{Cin} + \overline{A}\,\overline{B}Cin + \overline{A}B\overline{Cin} + A\overline{B}\,\overline{Cin} \tag{5.3}$$

A \ BCin	00	01	11	10
0			(1)	
1		(1)	(1)	(1)

Fig. 5.6 K-map for carry

$$\text{Carry} = AB + ACin + BCin \tag{5.4}$$

The expression of sum is further simplified using Boolean algebra laws and properties of logic gates.

$$\text{Sum} = AB\overline{Cin} + \overline{A}\,\overline{B}Cin + \overline{A}B\overline{Cin} + A\overline{B}\,\overline{Cin}$$

$$= Cin(AB + \overline{A}\,\overline{B}) + \overline{Cin}(A\overline{B} + \overline{A}B)$$

$$= Cin(A\ XNOR\ B) + \overline{Cin}(A\ XOR\ B)$$

$$= Cin(\overline{A\ XOR\ B}) + \overline{Cin}(A\ XOR\ B)$$

$$= Cin\ XOR\ (A\ XOR\ B) \qquad (A \oplus B = \overline{A}B + A\overline{B})$$

$$= Cin \oplus A \oplus B$$

So, $\text{Sum} = Cin \oplus A \oplus B$ $\hspace{2cm}$ (5.5)

The expression of carry is further simplified using Boolean algebra laws and properties of logic gates.

$$\text{Carry} = AB + ACin + BCin$$

$$= AB(Cin + \overline{Cin}) + ACin(B + \overline{B}) + BCin(A + \overline{A})$$

$$= ABCin + AB\overline{Cin} + ABCin + A\overline{B}Cin + ABCin + \overline{A}BCin$$

$$= ABCin + AB\overline{Cin} + A\overline{B}Cin + \overline{A}BCin \quad (A + A = A)$$

$$= AB(Cin + \overline{Cin}) + Cin(A \oplus B) \ (A + \overline{A} = 1)$$

$$= AB + Cin(A \oplus B)$$

So, Carry $= AB + Cin(A \oplus B)$ (5.6)

The simplified Boolean expression can be represented using logic gates as in fig. 5.7.

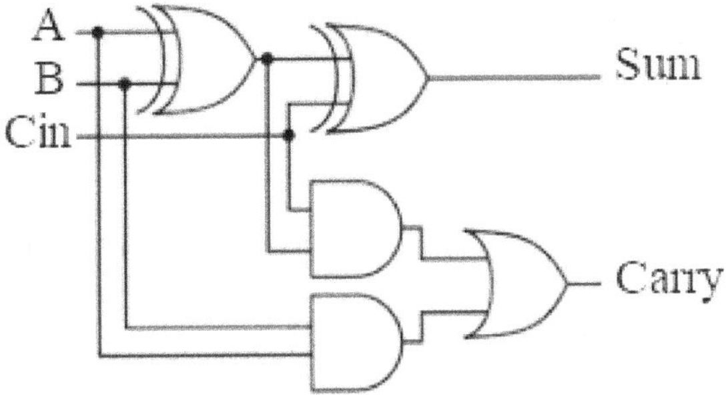

Fig. 5.7 Full Adder using logic gates

The full adder can be represented as the combination of two half adder as figure 5.8.

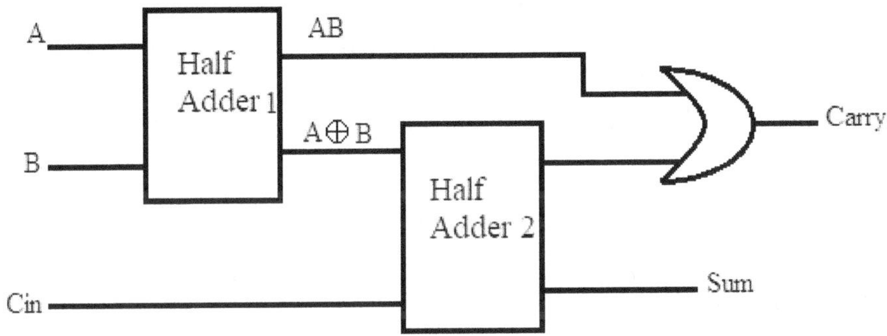

Fig. 5.8 Full adder using Half adder

Thus, the full adder is designed and verified.

EXPERIMENTAL PROCEDURE

1. For designing half adder circuit, make connections as per fig. 5.3.
2. Connect power supply to +5V to pin 14 and ground 0V to pin 7.
3. Connect two LEDs at output pins. One LED will represent output SUM and other LED will represent CARRY.
4. Apply various input conditions as per truth table 5.1. and verify the output.
5. Unwire the circuit.
6. Make connections for full adder as per fig. 5.7.
7. Connect power supply to +5V to pin 14 and ground 0V to pin 7.

8. Connect two LEDs at output pins. One LED will represent output SUM and other LED will represent CARRY.
9. Apply various input conditions as per truth table 5.2. and verify the output.
10. Full adder can be designed using two half adders, as per fig. 5.8.

RESULT

The truth table of half adder and full adder has been verified using logic gates.

PRECAUTIONS

1. Handle the ICs carefully.
2. Check the ICs before starting of the experiment.
3. The connections should be neat and tight.
4. The LED has one leg long and other leg short. The long leg depicts positive end and short end depicts negative end.
5. Do not press the IC on breadboard until pins are aligned with pours properly.
6. Avoid short circuit and heating of ICs.
7. Ground should be common.

VIVA VOCE QUESTIONS

1. What is the need of arithmetic circuits?
2. Design half adder and full adder using universal gates?
3. How binary adder is different than BCD adder?
4. What is the advantage of full adder over half adder?
5. Design BCD adder using universal logic gates?
6. What is the role of LED and resistor in the experiment?
7. What do you mean by parallel adder?

Binary Subtractor

OBJECTIVE

(A) To realize the Half Subtractor circuits using basic logic gates and to verify their truth tables.

(B) To realize the Full Subtractor circuits using basic logic gates and to verify their truth tables.

APPARATUS

Breadboard, Power supply (+5V), LEDs, Resistor 220Ω, Connecting wires.

IC REQUIRED

7486, 7408, 7404, 7400

INTRODUCTION

The subtractor is a digital logic circuits which implements subtraction of numbers. In combinational logic circuits, a subtractor performs the subtraction of two binary bits. Depending on the number of bits, the subtractors are further classified in two types as half subtractor and full subtractor.

HALF SUBTRACTOR

The half subtractor performs the operation of subtraction on two binary bits, minuend and subtrahend. It produces output in form of difference and borrow.

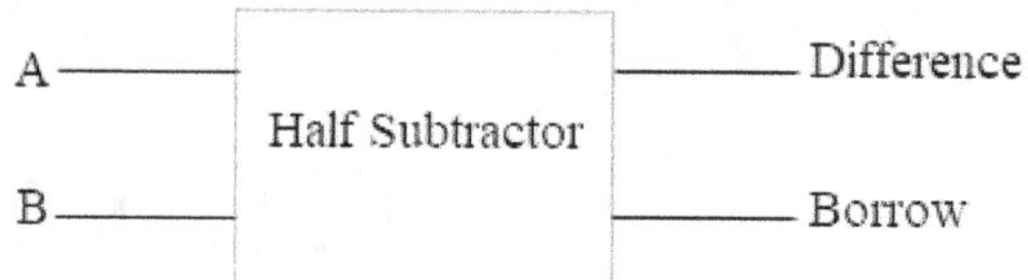

Fig. 6.1 Block diagram of Half Subtractor

The number of inputs and outputs are identified and labelled it. There are two inputs A and B producing two outputs Difference and Borrow. As the number of inputs are two, it suggests four possible combinations. The half subtractor performs binary subtraction. The fig. 6.1. shows the block diagram of half subtractor.

Table 6.1 Truth table of Half Subtractor

Inputs		Outputs	
A	**B**	**Difference**	**Borrow**
0	0	0	0
0	1	1	1
1	0	1	0
1	1	0	0

The table 6.1. shows the input-output relationship of half subtractor. A minimized Boolean function can be obtained using Karnaugh map. The two-variable K-map is used to explore the Boolean function.

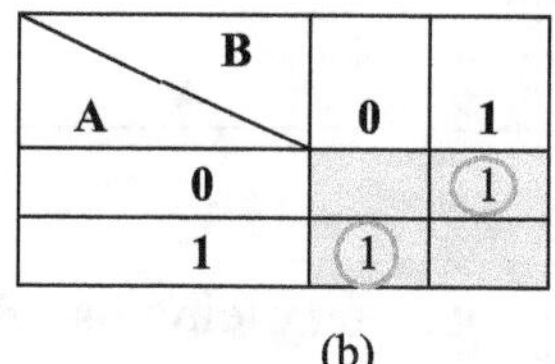

(a)   (b)

Fig. 6.2 K-map for (a) Difference and (b) Borrow

Using K-map, as in fig. 6.2., we get,

$$\text{Difference} = \bar{A}B \tag{6.1}$$

$$\text{Borrow} = A\bar{B} + \bar{A}B \tag{6.2}$$

The logical representation of Difference and Borrow using combination of EX-OR and AND gate, is shown as in fig. 6.3.

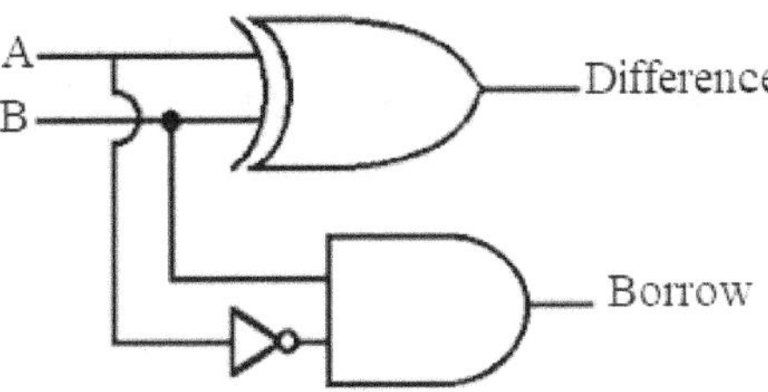

Fig. 6.3 Half Subtractor using logic gates

FULL SUBTRACTOR

In order to accommodate the borrow bit, generated by low order bits, the full subtractor performs the subtraction of three binary bits. It consists of three inputs A, B, B_{in} and two outputs difference, borrow.

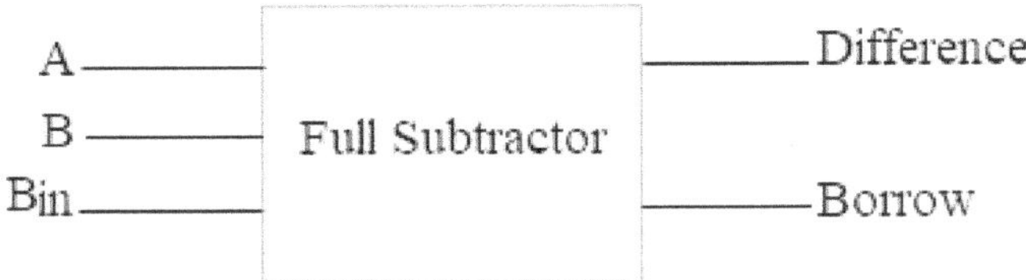

Fig. 6.4 Block diagram of Full Subtractor

The A and B are the significant binary bits and B_{in} is the borrow from low order bit. As the number of inputs are three, so the possible combinations are eight. The truth table is formulated which shows the output difference and borrow with respect to the all possible combinations of inputs. The block diagram of full subtractor is shown as in fig. 6.4. The table 6.2. shows the truth table of full subtractor.

Table 6.2 Truth table of full subtractor

Inputs			Outputs	
A	B	Bin	Difference	Borrow
0	0	0	0	0
0	0	1	1	1
0	1	0	1	1
0	1	1	0	1
1	0	0	1	0
1	0	1	0	0
1	1	0	0	0
1	1	1	1	1

A minimized Boolean function can be obtained using Karnaugh map. The three-variable K-map is used to explore the Boolean function, as in fig. 6.5.

A \ BBin	00	01	11	10
0		1		1
1	1		1	

Fig. 6.5(a) K-map for Difference

$$\text{Difference} = AB\overline{Bin} + \bar{A}\bar{B}Bin + \bar{A}B\overline{Bin} + A\bar{B}\overline{Bin} \tag{6.3}$$

A \ BBin	00	01	11	10
0		1	1	1
1			1	

Fig. 6.5(b) K-map for Borrow

$$\text{Borrow} = \bar{A}B + BBin + \bar{A}Bin \tag{6.4}$$

The expression of difference is further simplified using Boolean algebra laws and properties of logic gates.

$$\begin{aligned}
\text{Difference} &= AB\overline{Bin} + \bar{A}\bar{B}Bin + \bar{A}B\overline{Bin} + A\bar{B}\overline{Bin} \\
&= Bin(AB + \bar{A}\bar{B}) + \overline{Bin}(A\bar{B} + \bar{A}B) \\
&= Bin(A\,XNOR\,B) + \overline{Bin}(A\,XOR\,B) \\
&= Bin(\overline{A\,XOR\,B}) + \overline{Bin}(A\,XOR\,B) \\
&= Bin\,XOR\,(A\,XOR\,B) \qquad (A \oplus B = \bar{A}B + A\bar{B}) \\
&= Bin \oplus A \oplus B
\end{aligned}$$

So, $\text{Difference} = Bin \oplus A \oplus B \tag{6.5}$

The expression of borrow is further simplified using Boolean algebra laws and properties of logic gates.

$$\text{Borrow} = \bar{A}B + BBin + \bar{A}Bin$$

The simplified Boolean expression can be represented using logic gates as in fig. 6.6.

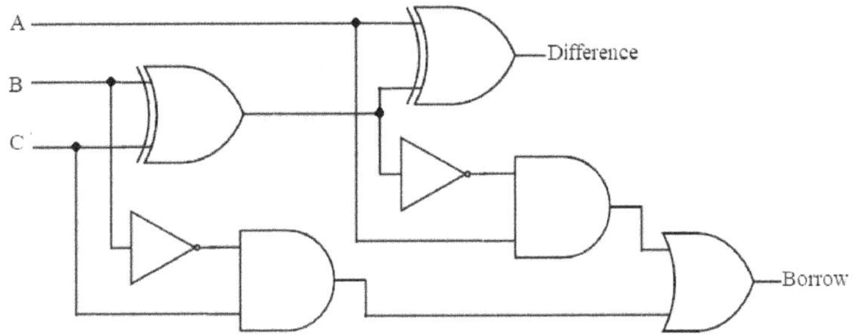

Fig. 6.6 Full subtractor using logic gates

FULL SUBTRACTOR USING HALF SUBTRACTOR

The full subtractor can be designed using half subtractor. The design is shown as in fig. 6.7. It consists of exclusive OR, AND, NOT & OR logic gates.

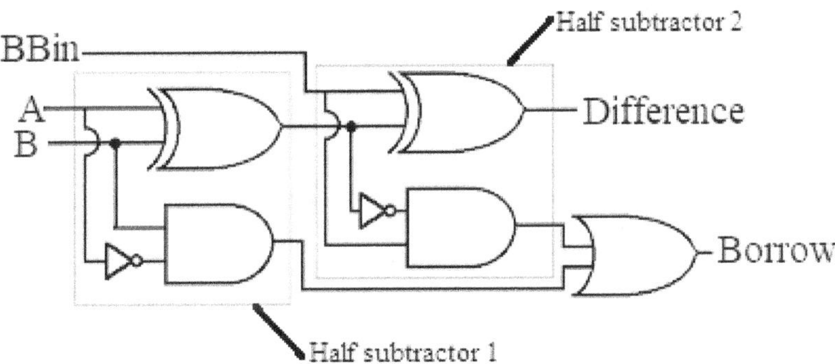

Fig. 6.7 Full subtractor using Half subtractor

The fig. 6.7. can be verified with the help of truth table 6.3.

Table 6.3 Truth table for full subtractor using half subtractor

Inputs			Outputs	
A	B	Bin	Difference	Borrow
0	0	0		
0	0	1		
0	1	0		
0	1	1		
1	0	0		
1	0	1		
1	1	0		
1	1	1		

EXPERIMENTAL PROCEDURE

1. For designing half adder circuit, make connections as per fig. 6.3.
2. Connect power supply to +5V to pin 14 and ground 0V to pin 7.
3. Connect two LEDs at output pins. One LED will represent output DIFFERENCE and another LED will represent BORROW.
4. Apply various input conditions as per truth table 6.1. and verify the output.
5. Unwire the circuit.
6. Make connections for full adder as per fig. 6.6.
7. Connect power supply to +5V to pin 14 and ground 0V to pin 7.
8. Connect two LEDs at output pins. One LED will represent output DIFFERENCE and another LED will represent BORROW.
9. Apply various input conditions as per truth table 6.2. and verify the output.
10. Full subtractor can be designed using two half subtractors, with the help of logic gates, as shown in fig. 6.7.
11. Truth table 6.3. can be verified using various combination of input variables.

RESULT

The truth table of half subtractor and full subtractor has been designed using logic gates, verified using logic gates.

PRECAUTIONS

1. Handle the ICs carefully.
2. Check the ICs before starting of the experiment.
3. The connections should be neat and tight.
4. The LED has one leg long and other leg short. The long leg depicts positive end and short end depicts negative end.
5. Do not press the IC on breadboard until pins are aligned with pours properly.
6. Avoid short circuit and heating of ICs.
7. Ground should be common.

VIVA VOCE QUESTIONS

1. What is the need and significance of complementary arithmetic?
2. Design half subtractor and full subtractor using universal gates?

3. What is the advantage of full subtractor over half subtractor?

4. How full subtractor is different than half subtractor?

5. Differentiate between combinational and sequential circuits?

6. Give the daily life examples of binary subtractor?

7. For 4 input variables A, B, C & D, how many combinations are possible?

Multiplexer

OBJECTIVE

(A)   To design 4:1 multiplexer using logic gates

(B)   To design 4:1 multiplexer using IC 74153

APPARATUS

Breadboard, Power supply (+5V), LEDs, Resistor 220Ω, Connecting wires.

IC REQUIRED

748153, 7432, 7404, 7411, 7408

INTRODUCTION

Multiplexer is a combinational logic circuit that is designed to choose one input out of various inputs. Here, 'various' refers to 2^n, n is number of data inputs. The function of multiplexer is controlled by select line or control signal. The shortened form of multiplexer is MUX or MPX. The multiplexer is also known as data selector, as it 'selects' input data, to be produced on output terminal.

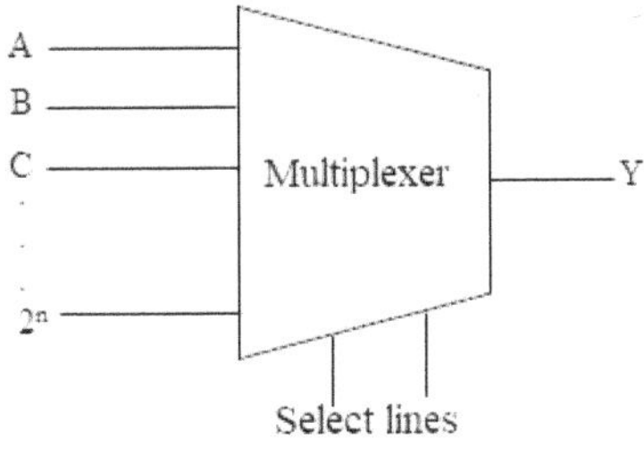

Fig. 7.1 Multiplexer

The fig. 7.1 shows the block diagram of multiplexer where number of inputs are 2^n and select lines are 'n'.

4:1 MULTIPLEXER

The 4 to 1 multiplexer, the 4 represents the number of inputs and 1 represents number of outputs. The number of select lines will be 1 ($2^2=4$; n=2). The fig.7.2. shows the configuration of 4:1 multiplexer.

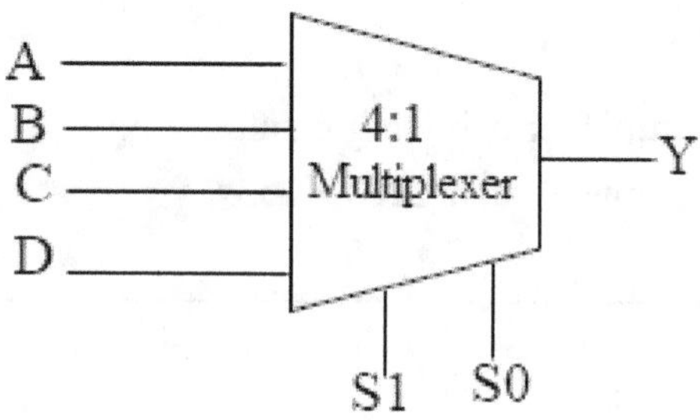

Fig. 7.2 4:1 multiplexer

The table 7.1. shows the truth table of 4 to 1 multiplexer. From the truth table, logical expression can be derived as:

Table 7.1 Truth table of 4:1 multiplexer

Select lines		Output
S1	**S0**	**Y**
0	0	A
0	1	B
1	0	C
1	1	D

The Select line 'S0 and S1' will decide which input will go to output terminal.

The output Y=A, when S1=0 and S0=0

The output Y=B, when S1=0 and S0=1

The output Y=C, when S1=1 and S0=0

The output Y=D, when S1=1 and S0=1

Therefore, $Y = A.\overline{S1}.\overline{S0} + B.\overline{S1}.S0 + C.S1.\overline{S0} + D.S1.S0$

The logic representation of 4:1 Multiplexer using logic gates is shown as in fig. 7.3.

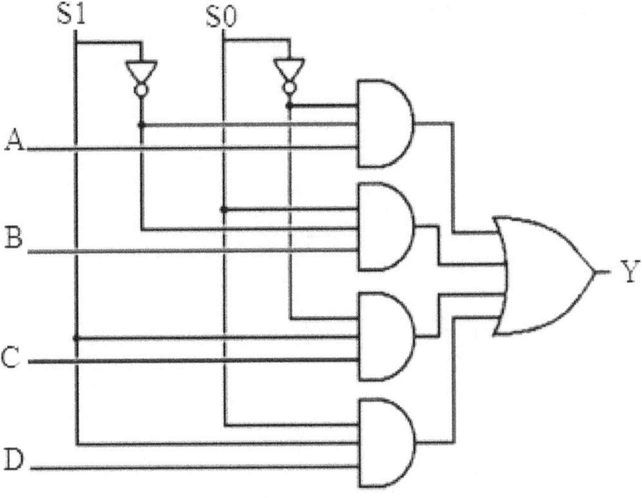

Fig. 7.3 Logic diagram of 4:1 multiplexer

IC 74153 PIN DIAGRAM

The IC 74153 contains two 4:1 multiplexer. Each multiplexer has strobe input G1 and G2 at pin 1 and pin 15 respectively. For valid output, the strobe input has to be at logic 0. The select lines are at pin 2 and pin 14. Two set of inputs can be given to IC 74153 i.e. pin 10-13 or pin 3-6. The output can be taken at pin 9 or pin 7, as per input set (fig. 7.4).

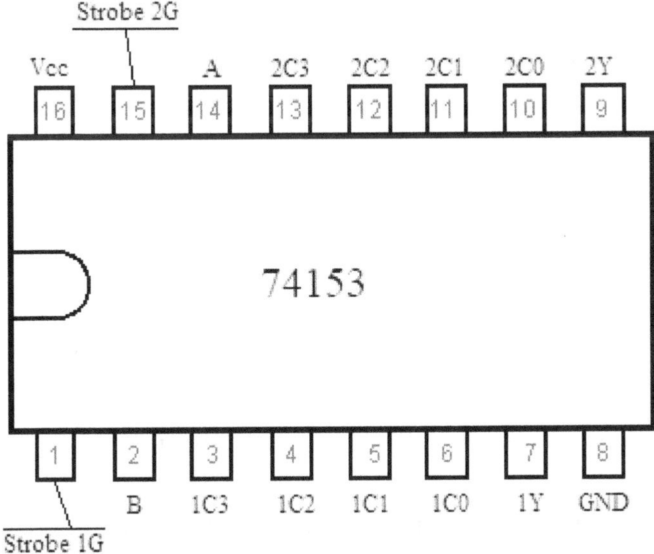

Fig. 7.4 Pin diagram of IC 74153

The truth table of two 4:1 multiplexer using IC74153 is shown as in table 7.2.

Table 7.2 4:1 multiplexer using IC74153

Strobe inputs		Select lines		Outputs	
G1	G0	S1	S0	IY	2Y
0	1	0	0	1C0	0
0	1	0	1	1C1	0
0	1	1	0	1C2	0
0	1	1	1	1C3	0
1	0	0	0	0	2C0
1	0	0	1	0	2C1
1	0	1	0	0	2C2
1	0	1	1	0	2C3

EXPERIMENTAL PROCEDURE

1. Check all the ICs using IC tester. After inserting IC carefully in ZIP socket of the tester, the tester should show "PASS".
2. Carefully insert the ICs in breadboard. Make connections as per fig. 7.3.
3. Connect the 5V (positive terminal) of power supply to pin 14 and 0V (negative terminal) to pin 7 of the IC.
4. Connect the LED through a resistor of 220Ω to output terminals of all OR gates.
5. Connect the circuit as per fig. 7.3 and verify the logic as per truth table 7.1. Thus, 4:1 multiplexer is designed using logic gates.
6. Now take IC74153, Connect +5V power supply to pin 16 and negative of power supply to pin 8.
7. Select input S0 and S1 on pin 2 and pin 14 and set them as 00. Strobe input G1G2 on pin 1 and pin 15 as 10. It implies that output data will be displayed at 2Y i.e. pin 9.
8. Connect the LEDs at pin 7 and pin 9.
9. Apply the inputs at pin 10, 11, 12 and 13. The input given at pin 10 will only be selected to appear at output pin 9.
10. Similarly, select the data for other input lines by verifying S1 S0 and G1 G2.
11. Verify the truth table as described in table 7.2.

RESULT

The truth table of 4:1 multiplexer is verified using

(a) logic gates (b) IC74153

PRECAUTIONS

1. Handle the ICs carefully.
2. Check the ICs before starting of the experiment. Test the ICs using digital IC tester.
3. The connections should be neat and tight.
4. The LED has one leg long and other leg short. The long leg depicts positive end and short end depicts negative end.
5. Do not press the IC on breadboard until pins are aligned with pours properly.
6. Avoid short circuit and heating of ICs.
7. Ground should be common.
8. For data input +5V must be given for HIGH and 0V for LOW.

VIVA VOCE QUESTIONS

1. What is the need of multiplexer?
2. Give any five daily life examples of multiplexer?
3. Design 4:1 multiplexer using 2:1 multiplexer
4. Why multiplexer is also known as data selector?
5. Design 4:1 multiplexer using NAND gate?
6. How can we use a multiplexer to implement a logic function?
7. Implement $F = \sum m(0, 1, 3, 5, 6, 7)$ using 4:1 multiplexer?

Demultiplexer

OBJECTIVE

(A) To design 1:2 demultiplexer using logic gates

(B) To design 1:4 demultiplexer using logic gates

(C) To design 1:8 demultiplexer using logic gates

APPARATUS

Breadboard, Power supply (+5V), LEDs, Resistor 220Ω, Connecting wires.

IC REQUIRED

7404, 7432, 7410, 7420, 7408

INTRODUCTION

The demultiplexer is a combinational logic circuit where one input signal is transmitting over various outputs. It receives data information on a single channel and distributes over several output lines. So, demultiplexer is a type of combinational logic circuit where number of inputs is single and number of outputs are 2^n. The demultiplexer is also known as data distributor. It is also known as serial to parallel converter.

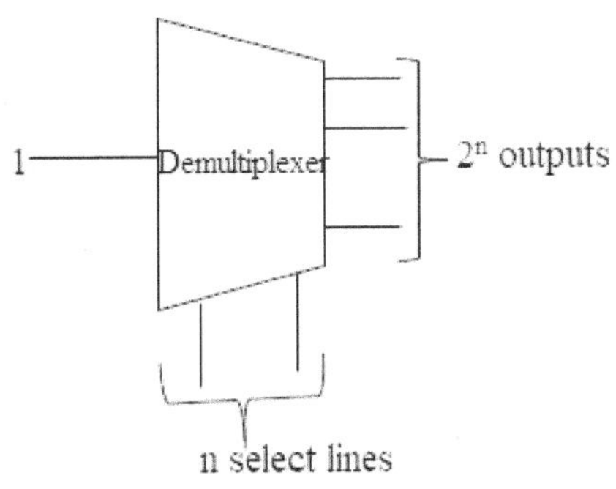

Fig. 8.1 demultiplexer

Figure 8.1 shows the block diagram of demultiplexer, where number of inputs is one, and outputs are 2^n.

1:2 DEMULTIPLEXER

A 1 to 2 demultiplexer has single input and 2 output lines (A and B). The number of select lines is 1 ($2=2^1$, select line=1). The table 8.1. shows the truth table of 1 to 2 demultiplexer.

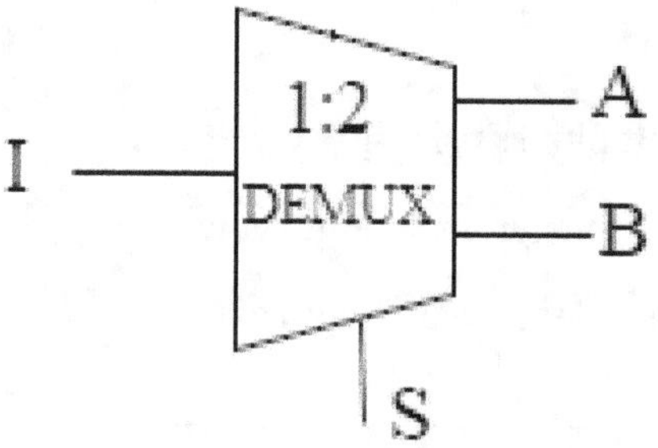

Fig. 8.2 1 to 2 demultiplexer

The truth table 8.1. depicts the functionality of 1 to 2 demultiplexer. The data input I is connected to output A, when the select line S is 0 and data input I is connected to output B, when the select line is 1(fig. 8.2).

Table 8.1 Truth table of 1 to 2 demultiplexer

Select line	Input	Output	
S	(I)	A	B
0	0	0	0
0	1	0	1
1	0	0	0
1	1	1	0

The expression for outputs can be written as:

$$A = I\bar{S}$$

$$B = IS$$

From these expressions, a 1 to 2 demultiplexer can be implemented using logic gates, as per Fig. 8.3.

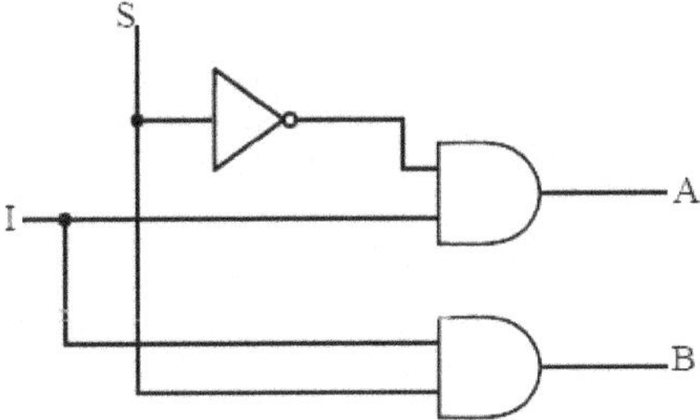

Fig. 8.3 Logic diagram of 1 to 2 demultiplexer

1:4 DEMULTIPLEXER

A 1 to 4 demultiplexer has single input and 4 output lines (A, B, C and D) (fig 8.4). The number of select lines are 2 ($4=2^2$, select line=2). The table 8.2. shows the truth table of 1 to 4 demultiplexer.

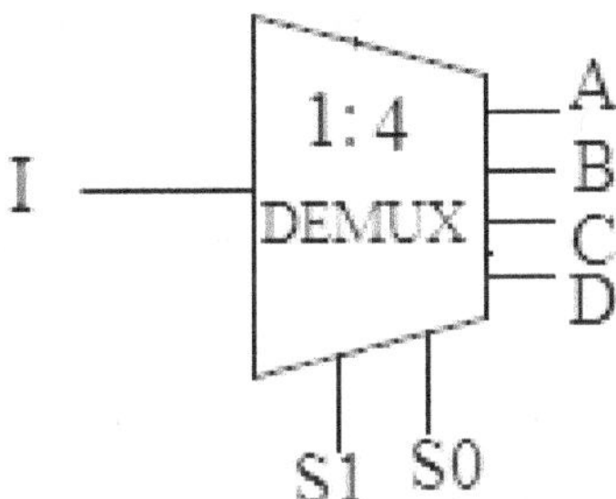

Fig. 8.4 1 to 4 demultiplexer

The truth table 8.2. depicts the functionality of 1 to 4 demultiplexer. The data input I is connected to output A, when the select line S1 and S0 are 0 and data input I is connected to output B, when the select line S1 is 0 and S0 is 1, the data input I is connected to output C, when select line S1 is 1 and S0 is 0 and data input I is connected to output D, when select lines S1 and S0 are 1. The demultiplexer will work, when enable signal is at logic 1 i.e. 'ON' mode.

Table 8.2 Truth table of 1 to 4 demultiplexer

Enable	Select line		Input	Output			
E	S1	S0		A	B	C	D
1	0	0	I	0	0	0	1
1	0	1	I	0	0	1	0
1	1	0	I	0	1	0	0
1	1	1	I	1	0	0	0

The expression for outputs can be written as:

$$A = I.\overline{S1}.\overline{S0}$$

$$B = I.\overline{S1}.S0$$

$$C = I.S1.\overline{S0}$$

$$D = I.S1.S0$$

From these expressions, a 1 to 4 demultiplexer can be implemented using logic gates, as per Fig. 8.5.

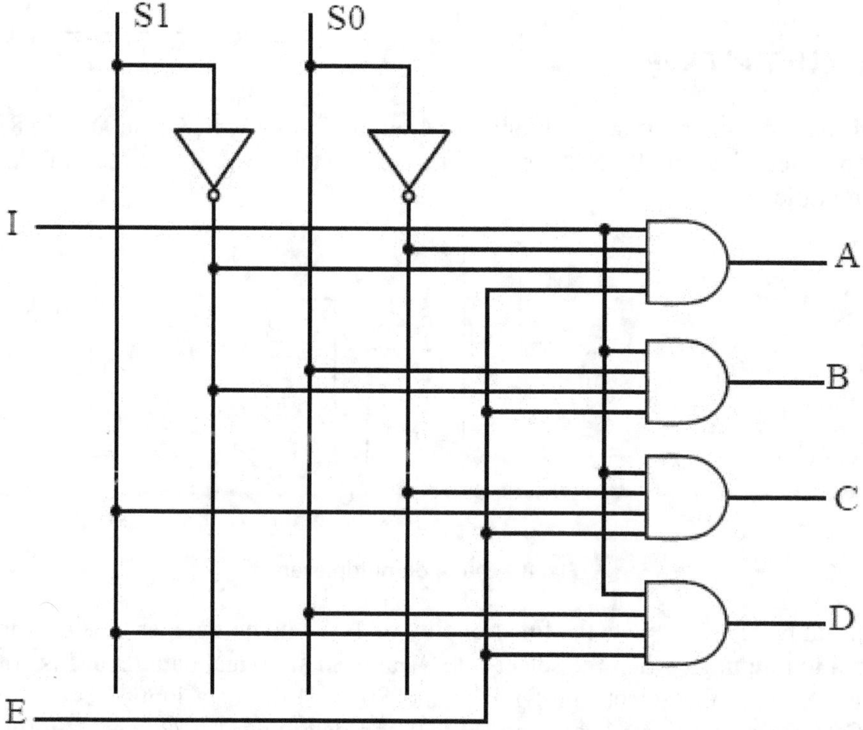

Fig. 8.5 Logic diagram of 1 to 4 demultiplexer

1:8 DEMULTIPLEXER

A 1 to 8 demultiplexer has single input and 8 output lines (A, B, C, D, E, F, G and H) (fig. 8.6). The number of select lines are 3 ($8=2^3$, select lines=3). The table 8.3. shows the truth table of 1 to 8 demultiplexer.

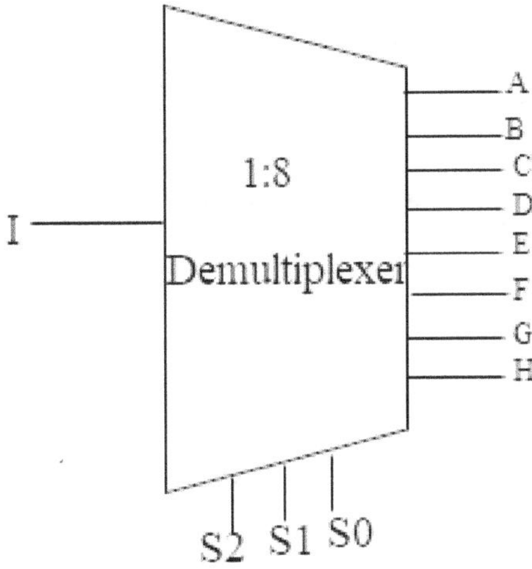

Fig. 8.6 1 to 8 demultiplexer

The truth table 8.3. depicts the functionality of 1 to 8 demultiplexer.

Table 8.3 Truth table of 1 to 8 demultiplexer

	Select line		Input	Output							
S2	**S1**	**S0**		**A**	**B**	**C**	**D**	**E**	**F**	**G**	**H**
0	0	0	I	0	0	0	0	0	0	0	1
0	0	1	I	0	0	0	0	0	0	1	0
0	1	0	I	0	0	0	0	0	1	0	0
0	1	1	I	0	0	0	0	1	0	0	0
1	0	0	I	0	0	0	1	0	0	0	0
1	0	1	I	0	0	1	0	0	0	0	0
1	1	0	I	0	1	0	0	0	0	0	0
1	1	1	I	1	0	0	0	0	0	0	0

Case 1: The data input I is connected to output A, when the select line S2, S1 and S0 are 0

Case 2: The data input I is connected to output B, when the select line S2 is 0, S1 is 0 and S0 is 1.

Case 3: The data input I is connected to output C, when select line S2 is 0, S1 is 1 and S0 is 0.

Case 4: The data input I is connected to output D, when select lines S2 is 0, S1 and S0 are 1.

Case 5: The data input I is connected to output E, when select line S2 is 1, S1 is 0 and S0 is 0

Case 6: The data input I is connected to output F, when select line S2 is 1, S1 is 0 and S0 is 1

Case 7: The data input I is connected to output G, when select line S2 is 1, S1 is 1 and S0 is 0

Case 8: The data input I is connected to output H, when select line S2 is 1, S1 is 1 and S0 is 1

The expression for outputs can be written as:

$$A = I.\overline{S2}.\overline{S1}.\overline{S0}$$

$$B = I.\overline{S2}.\overline{S1}.S0$$

$$C = I.\overline{S2}.S1.\overline{S0}$$

$$D = I.\overline{S2}.S1.S0$$

$$E = I.S2.\overline{S1}.\overline{S0}$$

$$F = I.S2.\overline{S1}.S0$$

$$G = I.S2.S1.\overline{S0}$$

$$H = I.S2.S1.S0$$

From these expressions, a 1 to 8 demultiplexer can be implemented using logic gates, as per Fig. 8.7.

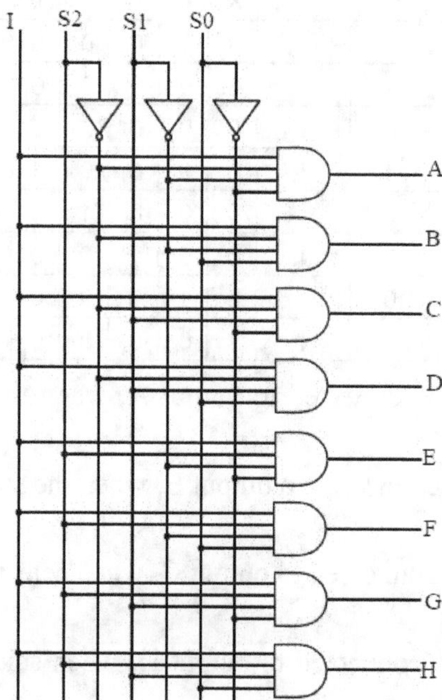

Fig. 8.7 Logic diagram of 1 to 8 demultiplexer

EXPERIMENTAL PROCEDURE

1. Check all the ICs using IC tester. After inserting IC carefully in ZIP socket of the tester, the tester should show "PASS".
2. Carefully insert the ICs in breadboard. Make connections as per fig. 8.3, 8.5 and 8.7.
3. Connect the 5V (positive terminal) of power supply to pin 14 and 0V (negative terminal) to pin 7 of the IC.
4. Connect the LED through a resistor of 220Ω to output terminals of all OR gates.
5. Connect the circuit as per fig. 8.3 for 1:2 demultiplexer, as fig. 8.5. for 1:4 demultiplexer and as fig. 8.7. for 1:8 demultiplexer.
6. Verify the logic as per truth table 8.1, 8.2 and 8.3 for 1:2 demultiplexer, 1:4 demultiplexer and 1:8 demultiplexer respectively. Thus, 4:1 multiplexer is designed using logic gates.

RESULT

(a) The truth table of 1:2 demultiplexer is verified using logic gates.

(b) The truth table of 1:4 demultiplexer is verified using logic gates.

(c) The truth table of 1:8 demultiplexer is verified using logic gates.

PRECAUTIONS

1. Handle the ICs carefully.
2. Check the ICs before starting of the experiment. Test the ICs using digital IC tester.
3. The connections should be neat and tight.
4. The LED has one leg long and other leg short. The long leg depicts positive end and short end depicts negative end.
5. Do not press the IC on breadboard until pins are aligned with pours properly.
6. Avoid short circuit and heating of ICs.
7. Ground should be common.
8. For data input +5V must be given for HIGH and 0V for LOW.

VIVA VOCE QUESTIONS

1. What is the need of demultiplexer?

2. Give any five daily life examples of demultiplexer?

3. Using IC74138, design 1:8 demultiplexer?

4. Design 1:4 demultiplexer using 1:2 demultiplexer?

5. Justify that demultiplexer acts as a data distributor?

6. Design 1:16 demultiplexer using 1:2 demultiplexer?

7. Name the IC for 1:2 demultiplexer, 1:4 demultiplexer, 1:8 demultiplexer and 1:16 demultiplexer.

Encoder

OBJECTIVE

(A)   To design Octal to Binary encoder using logic gates

(B)   To design Decimal to BCD encoder using logic gates

(C)   To design priority encoder using logic gates

APPARATUS

Breadboard, Power supply (+5V), LEDs, Resistor 220Ω, Connecting wires.

IC REQUIRED

7404, 7432, 7408, 7410

INTRODUCTION

Encoder is a combinational circuit that converts data information from 2^N input lines to N output lines. An encoder is used to convert analog signal into digital signal such as binary coded decimal form. The generalised block diagram of encoder is as shown in Fig. 9.1.

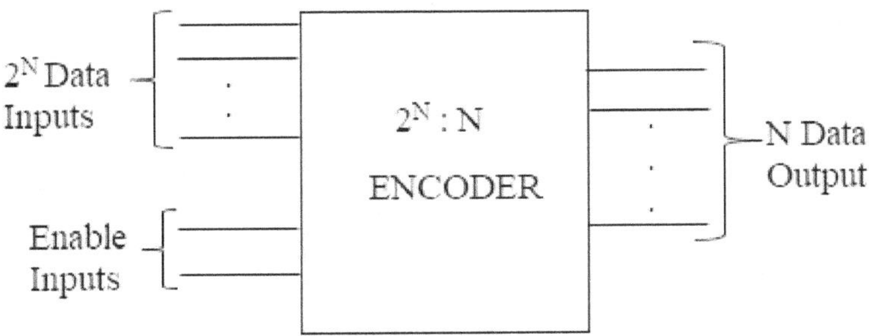

Fig. 9.1 Block diagram of an encoder

OCTAL TO BINARY ENCODER

An octal to binary encoder also known as 8 to 3 encoder, there will be eight inputs and three outputs (fig. 9.2).

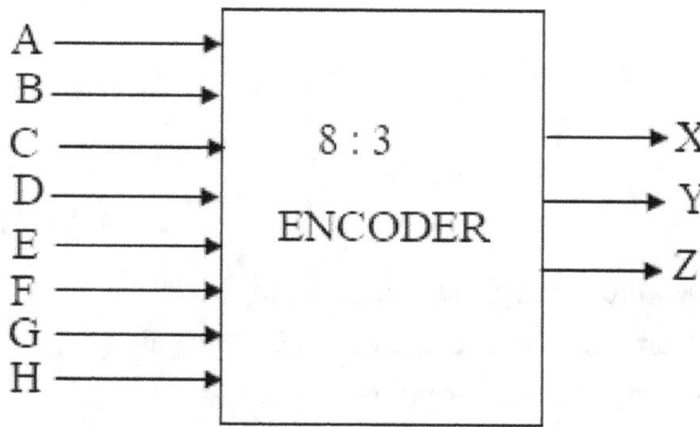

Fig. 9.2 A 8 to 3 encoder

The truth table 9.1. depicts the basic understanding of 8 to 3 encoder concept and output equations are framed.

Table 9.1 Truth table of octal to binary encoder

INPUTS								OUTPUTS		
A	B	C	D	E	F	G	H	X	Y	Z
1	0	0	0	0	0	0	0	0	0	0
0	1	0	0	0	0	0	0	0	0	1
0	0	1	0	0	0	0	0	0	1	0
0	0	0	1	0	0	0	0	0	1	1
0	0	0	0	1	0	0	0	1	0	0
0	0	0	0	0	1	0	0	1	0	1
0	0	0	0	0	0	1	0	1	1	0
0	0	0	0	0	0	0	1	1	1	1

$$Z = B + D + F + H$$

$$Y = C + D + G + H$$

$$X = E + F + G + H$$

From truth table, the output equations can be framed by looking into the input table. The placement of '1' in input table corresponding to '1' value of output, gives the expression for output equation.

$$X = E + F + G + H$$

$$Y = C + D + G + H$$

$$Z = B + D + F + H$$

The logical diagram of octal to binary encoder is represented as in Fig. 9.3.

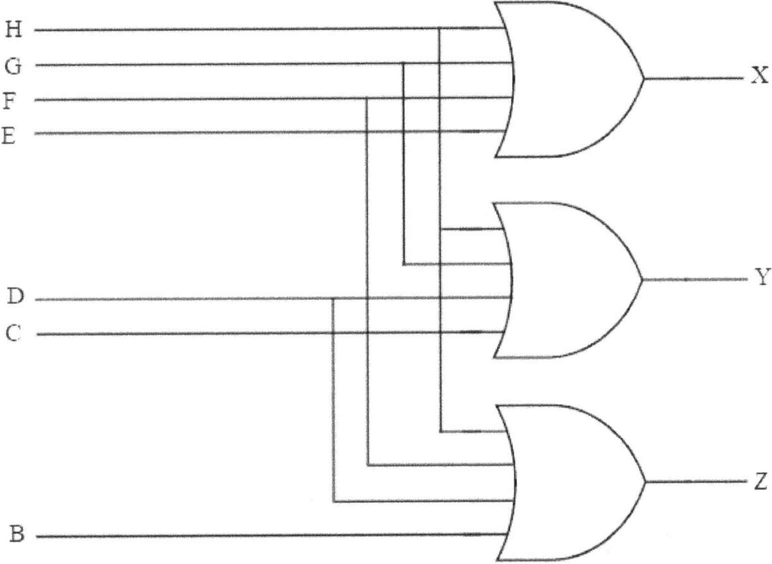

Fig. 9.3 Logic diagram of 8 to 3 encoder

DECIMAL TO BCD ENCODER

In decimal to BCD encoder, there will be 10 input lines and four will be selected as output lines (fig 9.4).

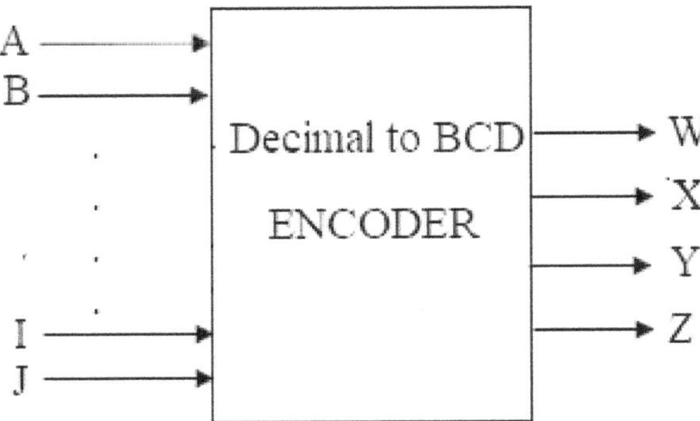

Fig. 9.4 Decimal to BCD encoder

The truth table 9.2. depicts the basic understanding of decimal to BCD encoder concept and output equations are framed.

Table 9.2 Truth table of decimal to BCD encoder

Decimal digit	BCD Code			
	A	B	C	D
0	0	0	0	0
1	0	0	0	1
2	0	0	1	0
3	0	0	1	1
4	0	1	0	0
5	0	1	0	1
6	0	1	1	0
7	0	1	1	1
8	1	0	0	0
9	1	0	0	1

From truth table, the output equations can be framed by looking into the input table. The corresponding value of input with respect to placement of '1' in output, gives the expression for output equation.

$$W = 8 + 9$$

$$X = 4 + 5 + 6 + 7$$

$$Y = 2 + 3 + 6 + 7$$

$$Z = 1 + 3 + 5 + 7 + 9$$

The logical diagram of decimal to BCD encoder is represented as in Fig. 9.5.

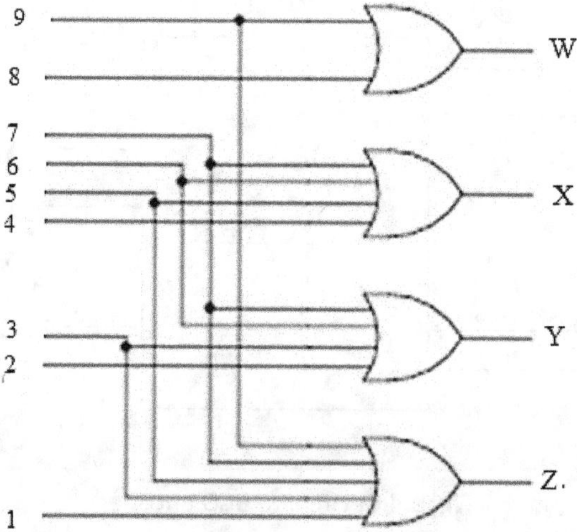

Fig. 9.5 Logic diagram of decimal to BCD encoder

PRIORITY ENCODER

The priority encoder is an encoder circuit which generates wrong code, in case more than one input is active. It works along with priority function. To resolve the problem of priority encoder, a level is assigned to each input, if more than one input is active at a same time, output will explore the levels of inputs and selects the input, whose subscript is having largest numerical value. It is used to compress, multiple inputs into the lesser number of outputs. The truth table for priority encoder is illustrated in table 9.3.

Table 9.3 Truth table of 4:2 priority encoder

INPUTS				OUTPUTS		
I_0	I_1	I_2	I_3	E	F	P
0	0	0	0	X	X	0
1	0	0	0	0	0	1
X	1	0	0	0	1	1
X	X	1	0	1	0	1
X	X	X	1	1	1	1

The 'X' designates an unknown number, it means the number may be '0' or '1'. The Input I_3 is having highest priority, so whenever I_3 is high, the output will be '1' irrespective of the values of other inputs. An indicator parameter 'P' is there, which accesses the priority level. If all inputs are '0', then the indicator 'P' is '0' and when one or more inputs are equal to '1', the indicator 'P' will be '1'. The expression of outputs can be explored using Karnaugh map (fig. 9.6).

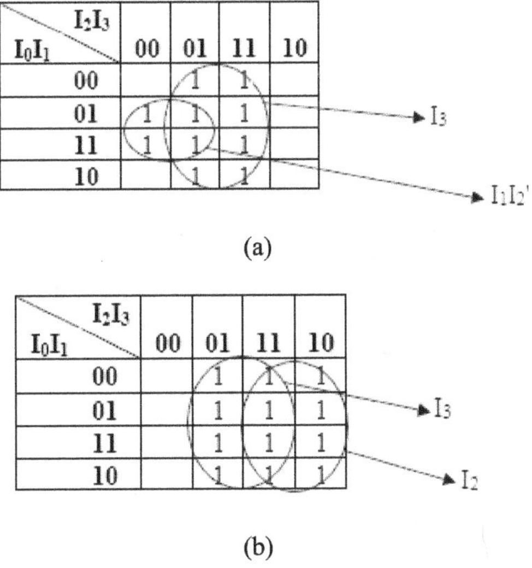

Fig. 9.6 K-map for output E (a) and F(b)

The expression for E, F and P are:

$$E = I3 + I1I2' \,,\, F = I2 + I3 \text{ and } P = I0 + I1 + I2 + I3$$

The logic diagram for priority encoder is (fig. 9.7).

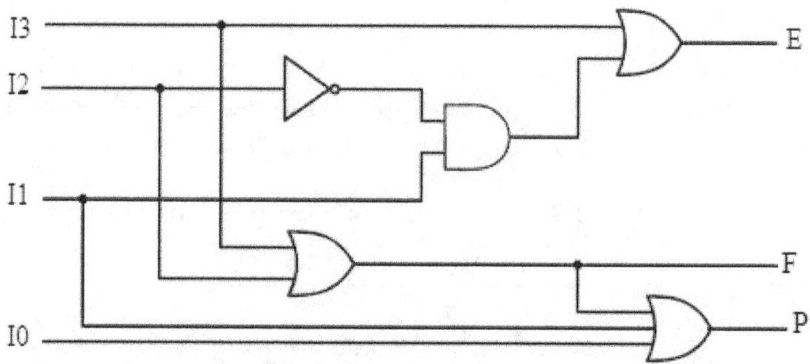

Fig. 9.7 Logic diagram for 4:2 Priority encoder

8 TO 3 PRIORITY ENCODER USING IC 74148

An IC 74148 is the most popularly used MSI encoder circuits for the 8 to 3 line priority encoder. The main characteristics of this encoder include cascading for priority encoding of n bits, code conversion, priority encoding of highest priority input line, decimal to BCD conversion, output enable-active low when all the inputs are high, etc.

The inputs of digital circuits often use octal code so there will be a need of entering such long binary words manually. Thus, the encoder IC is designed to achieve such operation.

The figure 9.8. shows its pin diagram and it has active low inputs and active low outputs. To handle more inputs, these ICs are cascaded by enabling input and Gray outputs which are also active low lines.

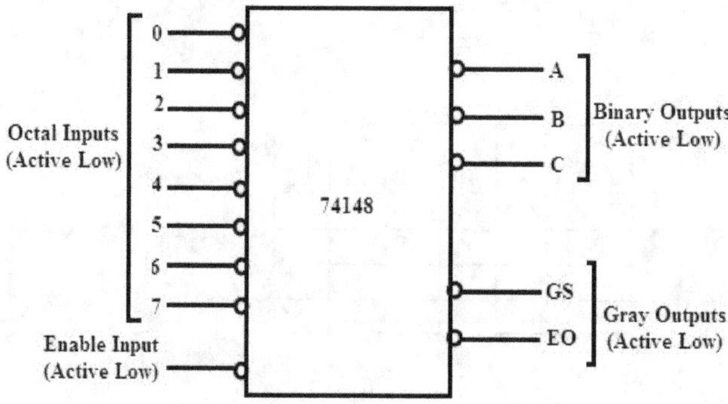

Fig. 9.8 Pin diagram of IC 74148

EXPERIMENTAL PROCEDURE

1. Check all the ICs using IC tester. After inserting IC carefully in ZIP socket of the tester, the tester should show "PASS".
2. Carefully insert the ICs in breadboard. Make connections as per fig. 9.3, 9.5 and 9.7.
3. Connect the 5V (positive terminal) of power supply to pin 14 and 0V (negative terminal) to pin 7 of the IC.
4. Connect the LED through a resistor of 220Ω to output terminals of all OR gates.
5. Connect the circuit as per fig. 9.3 for octal to binary, as fig. 9.5. for decimal to BCD and as fig. 9.7. for priority encoder.
6. Verify the logic as per truth table 9.1, 9.2 and 9.3 for octal to binary, decimal to BCD, priority encoder respectively. Thus, various forms of encoder are designed using logic gates.
7. Octal to binary encoder is verified using IC 74148, where octal inputs are given at respective input pins and outputs are observed at output pins.

RESULT

(a)   The truth table of octal to binary encoder is verified using logic gates.

(b)   The truth table of decimal to BCD encoder is verified using logic gates.

(c)   The truth table of priority encoder is verified using logic gates.

PRECAUTIONS

1. Handle the ICs carefully.
2. Check the ICs before starting of the experiment. Test the ICs using digital IC tester.
3. The connections should be neat and tight.
4. The LED has one leg long and other leg short. The long leg depicts positive end and short end depicts negative end.
5. Do not press the IC on breadboard until pins are aligned with pours properly.
6. Avoid short circuit and heating of ICs.
7. Ground should be common.
8. For data input +5V must be given for HIGH and 0V for LOW.

VIVA VOCE QUESTIONS

1. What is the need of encoder?
2. Give any five daily life examples of encoder?
3. Name the IC for all variants of encoder?
4. Give the applications of priority encoder?
5. Design 16 to 4 priority encoder?
6. What is the significance of 'priority' in priority encoder?
7. How encoders are different than multiplexers?

Decoder

OBJECTIVE

(A)　To design 2 line to 4 line decoder using logic gates

(B)　To design 3 line to 8 line decoder using logic gates

(C)　To design 3 line to 8 line decoder using IC 74138

APPARATUS

Breadboard, Power supply (+5V), LEDs, Resistor 220Ω, Connecting wires.

IC REQUIRED

7404, 7432, 7408 (2-input AND gate), 7411 (3-Input AND gate), 74138

INTRODUCTION

A n-bit binary code has distinct 2^n units. The decoder is a combinational circuit that detects a particular digital state. It is used to convert n input binary numbers to 2^n output binary numbers, here, n is number of bits. The block diagram of decoder is as shown in fig. 10.1.

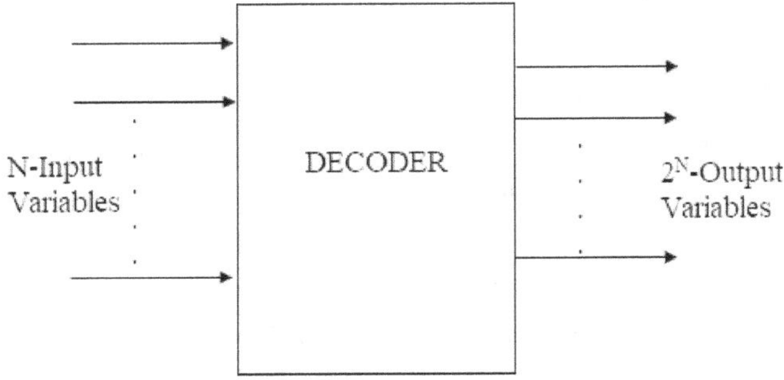

Fig. 10.1 Block diagram of decoder

2 LINE TO 4 LINE DECODER

In 2 to 4 decoder, the number of input variables are two and output variables are $2^n=2^2=4$. The block diagram of two to four decoder is shown as in fig. 10.2.

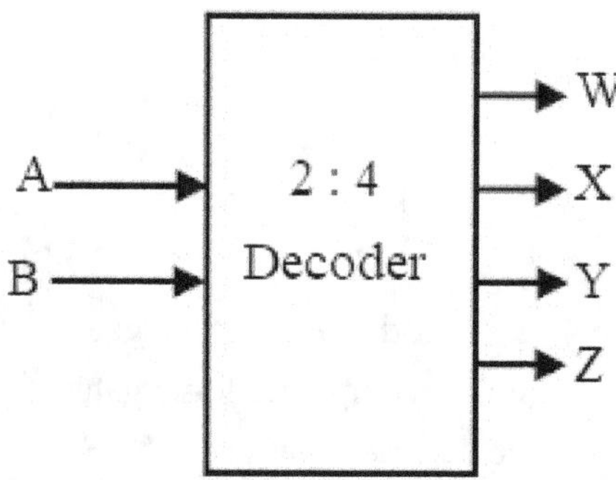

Fig. 10.2 Block diagram of 2 to 4 decoder

The truth table of 2 to 4 decoder is depicted in table 10.1., where inputs and outputs are displayed. The decoder will work when enable line is active (logic 1).

Table 10.1 Truth table of 2 to 4 decoder

Inputs			Outputs			
E	A	B	W	X	Y	Z
0	X	X	0	0	0	0
1	0	0	0	0	0	1
1	0	1	0	0	1	0
1	1	0	0	1	0	0
1	1	1	1	0	0	0

When the enable is '0', irrespective of the value of input A and B, output is always '0'. The decoder will work, when the enable is '1'. The output 'Z' will active, when the input is AB=0, the output 'Y' will active, when the input is AB=01, similarly when the input AB is 10 or 11, the output 'X' or 'W' will be activated respectively. A decoder will enable input can function like a demultiplexer, with the difference that number of inputs is 1 in case of demultiplexer. The logic diagram for 2 to 4 decoder is shown as in fig. 10.3.

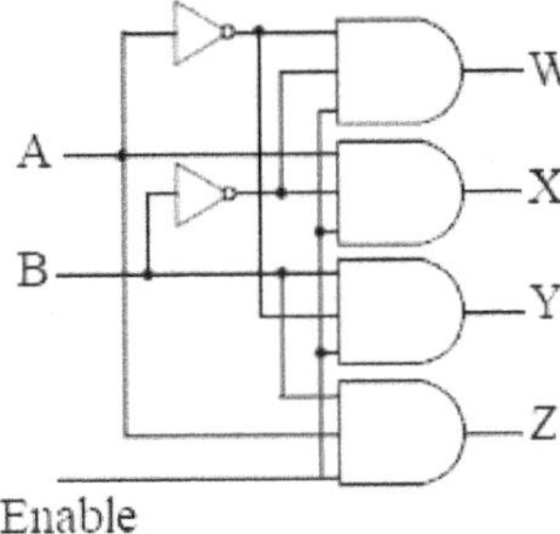

Fig. 10.3 Logic diagram for 2 to 4 decoder

---

3 LINE TO 8 LINE DECODER

In 3 to 8 decoder, the number of input variables are three and output variables are $2^n=2^3=8$. The block diagram of three to eight decoder is shown as in fig. 10.4.

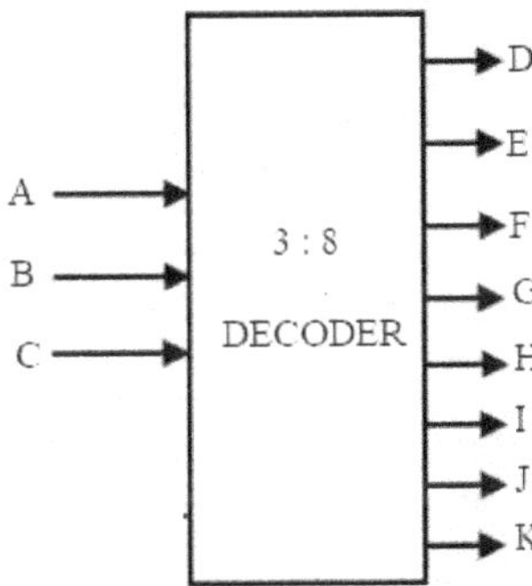

Fig. 10.4 Block diagram of 3 to 8 decoder

The truth table of 3 to 8 decoder is depicted in table 10.2., where inputs and outputs are displayed. The decoder will work when enable line is active (logic 1).

Table 10.2 Truth table of 3 to 8 decoder

Inputs			Outputs							
A	B	C	D	E	F	G	H	I	J	K
0	0	0	1	0	0	0	0	0	0	0
0	0	1	0	1	0	0	0	0	0	0
0	1	0	0	0	1	0	0	0	0	0
0	1	1	0	0	0	1	0	0	0	0
1	0	0	0	0	0	0	1	0	0	0
1	0	1	0	0	0	0	0	1	0	0
1	1	0	0	0	0	0	0	0	1	0
1	1	1	0	0	0	0	0	0	0	1

When the enable is '0', irrespective of the value of input A, B and C, output is always '0'. The decoder will work, when the enable is '1'. The output 'D' will active, when the input is ABC=000, the output 'E' will active, when the input is ABC=001, similarly when the input ABC is 010, output 'F' will active, when ABC=011, the output 'G', at ABC=100, the output 'H', at ABC=101, the output 'I' and at input ABC=110, the output 'J' will be activated respectively. At input ABC=111, the output 'K' will be high. A decoder will enable input can function like a demultiplexer, with the difference that number of inputs is 1 in case of demultiplexer. The logic diagram for 3 to 8 decoder is shown as in fig. 10.5.

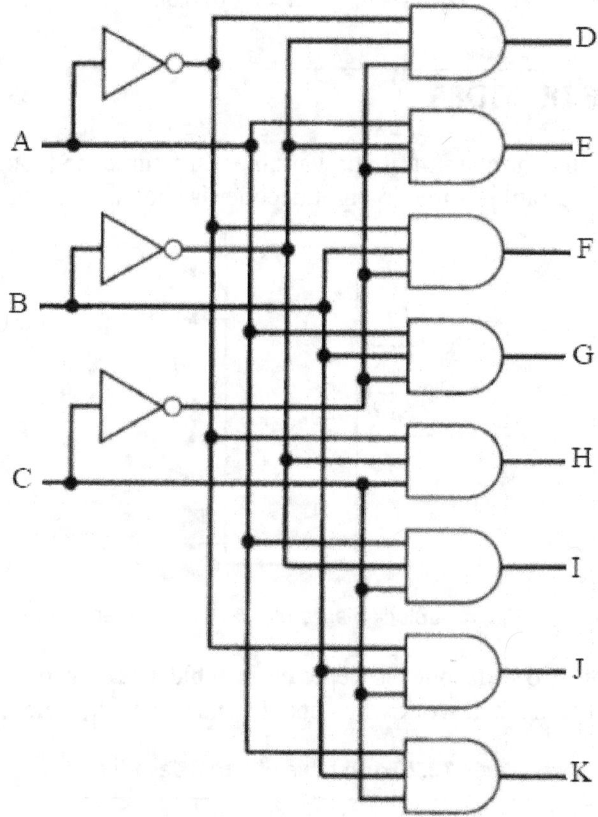

Fig. 10.5 Logic diagram for 3 to 8 decoder

3 LINE TO 8 LINE DECODER USING IC 74138

The 3 line to 8 line decoder can also be designed using IC74138. This is 16 pin IC. It has 3 inputs and 8 output pins. The pin 16 is reserved for Vcc and pin 8 is used for ground purpose. There are 3 enable pins for control the operation. If the enable pins are not activated, then all the eight outputs will be high irrespective of the value of inputs.

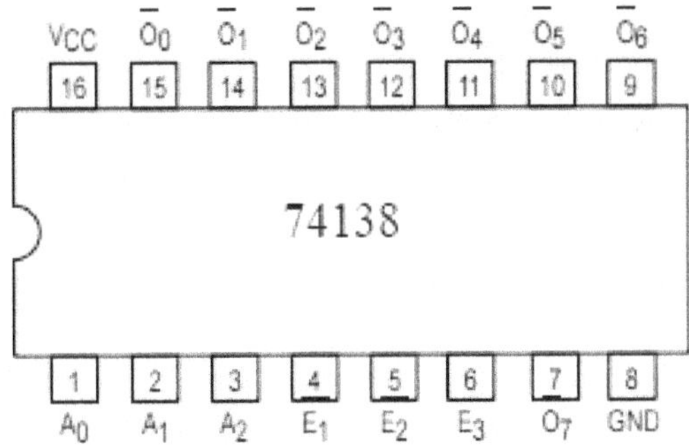

Fig. 10.6 3 Line to 8 Line Decoder

The figure 10.6. shows the pin diagram of IC74138 along with its input and output pins labels.

Table 10.3 Logic verification using IC74138

Inputs			Outputs							
A0	A2	A3	O0	O1	O2	O3	O4	O5	O6	O7
0	0	0								
0	0	1								
0	1	0								
0	1	1								
1	0	0								
1	0	1								
1	1	0								
1	1	1								

While putting the different values of A0, A1 and A2, the respective outputs O0-O7 are observed at output terminals.

EXPERIMENTAL PROCEDURE

1. Check all the ICs using IC tester. After inserting IC carefully in ZIP socket of the tester, the tester should show "PASS".

2. Carefully insert the ICs in breadboard. Make connections as per fig. 10.3 and 10.5.

3. Connect the 5V (positive terminal) of power supply to pin 14 and 0V (negative terminal) to pin 7 of the IC.

4. Connect the LED through a resistor of 220Ω to output terminals of all OR gates.

5. Connect the circuit as per fig. 10.3 for 2 line to 4 line decoder and as fig.10.5. for 3 line to 8 line decoder.

6. Verify the logic as per truth table 10.1 and 10.2. for 2 line to 4 line decoder and 3 line to 8 line decoder respectively. Thus, various forms of decoder are designed using logic gates.

7. Unwire the circuit connections and take IC74138.

8. 3 line to 8 line decoder is verified using IC 74138, where three inputs are given at respective input pins and outputs are observed at output pins.

9. Verify the table 10.3. for various set of input combinations and observe the output values.

RESULT

(a) The truth table of 2 line to 4 line decoder is verified using logic gates.

(b) The truth table of 3 line to 8 line decoder is verified using logic gates.

(c) The truth table of 3 line to 8 line decoder is verified using IC 74138.

PRECAUTIONS

1. Handle the ICs and electronic components carefully.

2. Check the ICs before starting of the experiment. Test the ICs using digital IC tester.

3. The connections should be neat and tight.

4. The LED has one leg long and other leg short. The long leg depicts positive end and short end depicts negative end.

5. Do not press the IC on breadboard until pins are aligned with pours properly.

6. Avoid short circuit and heating of ICs.

7. Ground should be common.

8. For data input +5V must be given for HIGH and 0V for LOW.

VIVA VOCE QUESTIONS

1. What is the need of decoder?

2. Give any five daily life examples of decoder?

3. Name the IC for all variants of decoder?

4. Differentiate between demultiplexer and decoder?

5. Design 4 line to 16 line decoder using logic gates?

6. Design 5 line to 32 line decoder using 2 line to 4 line and 3 line to 8 line decoder?

7. Design 4 line to 16 line decoder using 3 line to 8 line decoder?

Magnitude Comparator

OBJECTIVE

(A) To design a magnitude comparator to compare two 1-bit numbers using logic gates.

(B) To design a magnitude comparator to compare two 2-bit numbers using logic gates.

(C) To design magnitude comparator to compare two 4-bit numbers using IC 7485.

APPARATUS

Breadboard, Power supply (+5V), LEDs, Resistor 220Ω, Connecting wires.

IC REQUIRED

7404,7486, 7432, 7408 (2-input AND gate), 7411 (3-Input AND gate), 7485 (4-bit magnitude comparator)

INTRODUCTION

Digital comparator is used to compare the digital signals at input terminal and produces an output, as per the outcome of comparison. Mainly, two types of digital comparators are there to compare binary bits.

(a) **Identity comparator:** This is one type of digital comparator, where the output terminal is single. For example, A and B are two binary numbers, whose identity needs to compare. It will compare A with B, for condition A=B and produces output 1 (true) (A=B=0 or A=B=1) or 0 (false) (A $\neq$ B), as per the comparison output. Only one type of comparison (A=B) is done with this comparator.

(b) **Magnitude comparator:** In this type of digital comparator, there are three output terminals. It will compare binary bits A and B for conditions equal to, greater than and less than i.e. A=B, A<B and A>B. Three types of comparisons are performed using magnitude comparator. The general specification of n-bit comparator is shown in fig. 11.1.

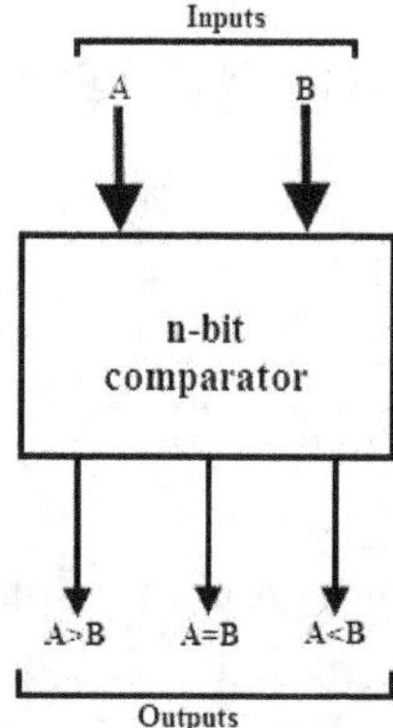

Fig. 11.1 n-bit comparator

Basically, digital comparators use exclusive-NOR gate in their design and operation, because comparator will generate output 1, when two inputs are equal.

ONE BIT MAGNITUDE COMPARATOR

The one-bit magnitude comparator performs its operation when input consists of one bit. For example, A and B are two one-bit binary number, comparator will check all the three conditions: equal to, greater than and less than.

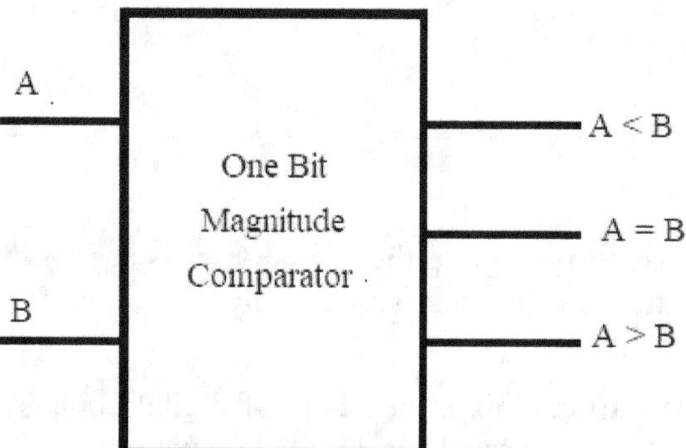

Fig. 11.2 Block diagram of one-bit magnitude comparator

The fig. 11.2. represents the block diagram of one-bit magnitude comparator. As the number of inputs are two, the possible combinations will be four. The truth table 11.1. depicts the comparison of A and B.

Table 11.1 One-bit magnitude comparator

Inputs		Outputs		
A	B	A<B	A=B	A>B
0	0	0	1	0
0	1	1	0	0
1	0	0	0	1
1	1	0	1	0

ONE BIT MAGNITUDE COMPARATOR USING LOGIC GATES

The Boolean expression can be explored by using Karnaugh map. There will be three K-maps for output, as shown in fig. 11.3.

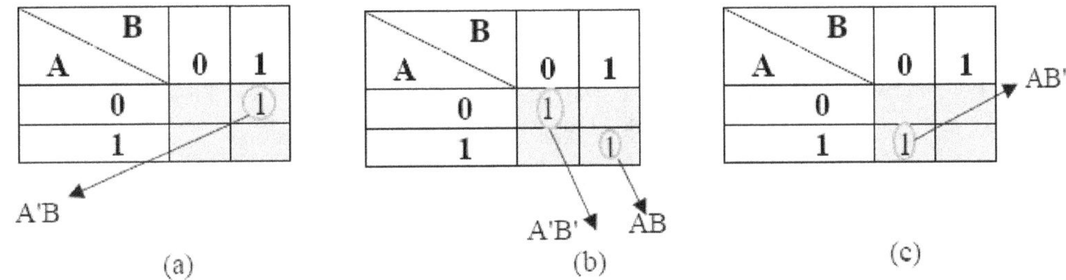

Fig. 11.3 K-map for (a) A<B (b) A=B (c) A>B

So, Boolean expression for $< B = \bar{A}B$, $A = B$ is $\bar{A}B + AB$ and A>B is $A\bar{B}$.

The logic diagram for one-bit magnitude comparator is shown as in fig. 11.4.

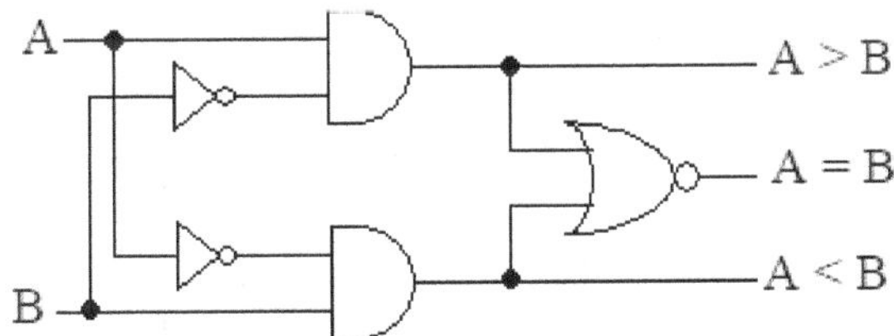

Fig. 11.4 Logic diagram of one-bit magnitude comparator

TWO BIT MAGNITUDE COMPARATOR

The two-bit magnitude comparator performs its operation when input consists of two bits. For example, A and B are two-bit binary numbers, comparator will check all the three conditions: equal to, greater than and less than.

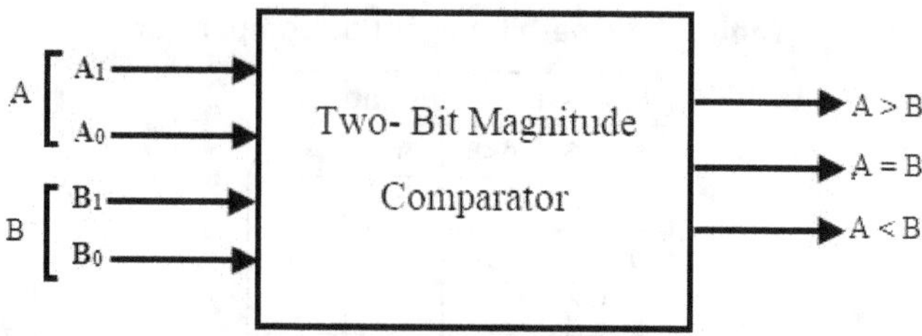

Fig. 11.5 Block diagram of one-bit magnitude comparator

The fig. 11.5. represents the block diagram of one-bit magnitude comparator. As each binary number consists of two bits, all the possible combinations are shown in table 11.2.

Table 11.2 Two-bit magnitude comparator

Inputs				Outputs		
A		**B**				
A0	**A1**	**B0**	**B1**	**A<B**	**A=B**	**A>B**
0	0	0	0	0	1	0
0	0	0	1	1	0	0
0	0	1	0	1	0	0
0	0	1	1	1	0	0
0	1	0	0	0	0	1
0	1	0	1	0	1	0
0	1	1	0	1	0	0
0	1	1	1	1	0	0
1	0	0	0	0	0	1
1	0	0	1	0	0	1
1	0	1	0	0	1	0
1	0	1	1	1	0	0
1	1	0	0	0	0	1
1	1	0	1	0	0	1
1	1	1	0	0	0	1
1	1	1	1	0	1	0

TWO BIT MAGNITUDE COMPARATOR USING LOGIC GATES

The Boolean expression can be explored by using Karnaugh map. There will be three K-maps for output, as shown in fig. 11.6.

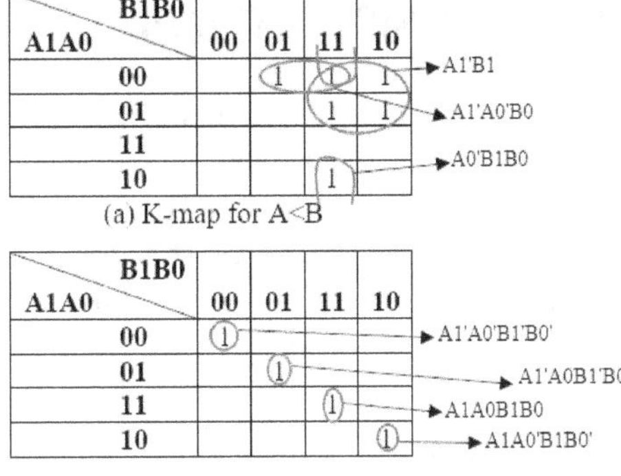

(a) K-map for A<B

(b) K-map for A=B

(c) K-map for A>B

Fig. 11.6 K-map for (a) A<B (b) A=B (c) A>B

The logic diagram for two-bit magnitude comparator is shown as in fig. 11.7.

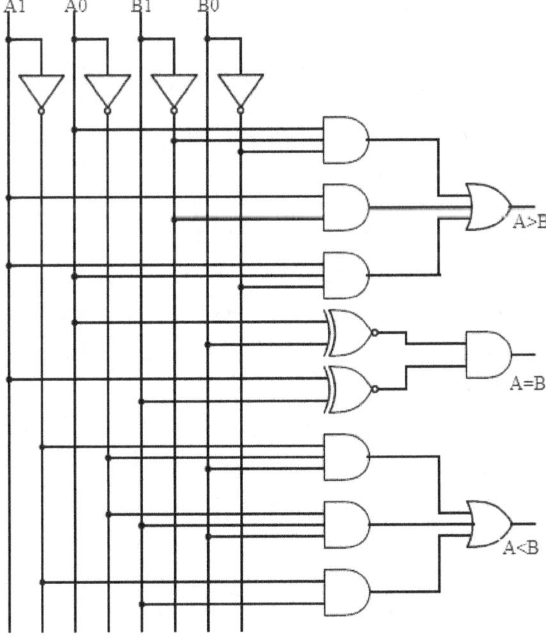

Fig. 11.7 Logic diagram of two-bit magnitude comparator

FOUR BIT MAGNITUDE COMPARATOR USING IC7485

The two 4-bit magnitude comparator can be designed using IC7485. It is a 4-bit magnitude comparator with three cascaded inputs. The comparison can be extended to any number of bits by cascading the comparators. To expand the comparison the A>B, A=B and A<B output pins (i.e. pin 5, 6 and 7) of the lower significant comparator are connected to the corresponding cascading inputs of next higher comparator i.e. connected to pin 4, 3 and 2 (A>B, A=B and A<B)

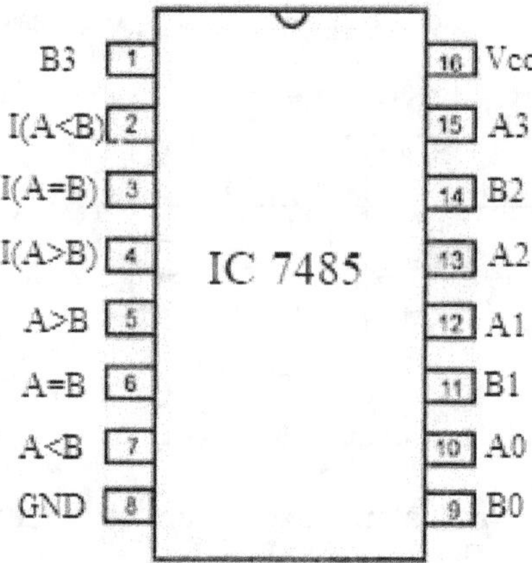

Fig. 11.8 Pin diagram of IC7485

The pin diagram of magnitude comparator IC7485 is shown as in fig. 11.8.

EXPERIMENTAL PROCEDURE

1. Check all the ICs using IC tester. After inserting IC carefully in ZIP socket of the tester, the tester should show "PASS".
2. Carefully insert the ICs in breadboard. Make connections as per fig. 11.4 and 11.7.
3. Connect the 5V (positive terminal) of power supply to pin 14 and 0V (negative terminal) to pin 7 of the IC.
4. Connect the LED through a resistor of 220Ω to output terminals of all OR gates.
5. Connect the circuit as per fig. 11.4. for one-bit magnitude comparator and as fig.11.7. for two-bit magnitude comparator.
6. Verify the logic as per truth table 11.1 and 11.2. for one-bit magnitude comparator and two-bit magnitude comparator. Thus, various forms of magnitude comparator are designed using logic gates.
7. Unwire the circuit connections and take IC7485.
8. Connect 5V power supply at pin 16 and ground to pin 8 of IC7485.

9. Connect the first 4-bit binary input (A) to pin 10,12,13 and 15 with pin 10 being least significant digit and 15 being most significant digit.

10. Connect the second 4-bit binary input (B) to pin 9,11,14 and 1 with pin 9 being least significant digit and 1 being most significant digit.

11. Connect LED at pin 5, 6 and 7. Pin 5 corresponds to A>B, Pin 6 corresponds to A=B and pin 7 corresponds to A<B. Only one LED should glow at a time.

RESULT

(A) A magnitude comparator is designed to compare two 1-bit numbers using logic gates.

(B) A magnitude comparator is designed to compare two 2-bit numbers using logic gates.

(C) A magnitude comparator is designed to compare two 4-bit numbers using IC 7485.

PRECAUTIONS

1. Handle the ICs and electronic components carefully.
2. Check the ICs before starting of the experiment. Test the ICs using digital IC tester.
3. The connections should be neat and tight.
4. The LED has one leg long and other leg short. The long leg depicts positive end and short end depicts negative end.
5. Do not press the IC on breadboard until pins are aligned with pours properly.
6. Avoid short circuit and heating of ICs.
7. Ground should be common.
8. For data input +5V must be given for HIGH and 0V for LOW.

VIVA VOCE QUESTIONS

1. What is the need of comparator?
2. Give any five daily life examples of comparator?
3. Which gate is known as basic comparator and why?
4. How many IC7485 are needed to compare two 8-bit numbers?
5. Can NOR gate be used to compare two binary numbers?
6. Why do we say one binary number differs than the other binary number?
7. Give the pin specification of IC7485?

Flip-Flops

OBJECTIVE

(A)   To design SR flip-flop using NAND gates.

(B)   To design JK flip-flop using NAND gates.

(C)   To design Master-Slave JK flip-flop using NAND gates.

(C)   To design D flip-flop using NAND gates.

(D)   To design T flip-flop using NAND gates.

APPARATUS

Breadboard, Power supply (+5V), LEDs, Resistor 220Ω, Connecting wires.

IC REQUIRED

7400, 7404, 7410 (3 input NAND gate)

INTRODUCTION

A latch is an asynchronous sequential circuit, where any change in input can be seen on the output immediately. Further, to modify the operation of latch, an extra control signal is associated with latch which controls the transition of input signal. This extra control signal is called clock or clock pulse. A latch with clock is known as flip-flop.

Flip-flop is also known as bi-stable latch because when the circuit is triggered by input pulse, it will remain in that state unless until further input pulse is not applied. So, the circuit remembers that state, so it is also known as bi-stable latch. On the basis of operation and process, there are four different types of flip-flops:

1. SR flip-flop

2. JK flip-flop

3. D flip-flop

4. T flip-flop

SR FLIP-FLOP

The SR flip-flops are also known as 1-bit memory because even after the input pulse, the value of input pulse is stored in SR flip-flop. The circuit will work only when clock is activated. Further, its working is based on positive edge triggering or negative edge triggering. The symbol of SR flip-flop is shown as in fig. 12.1.

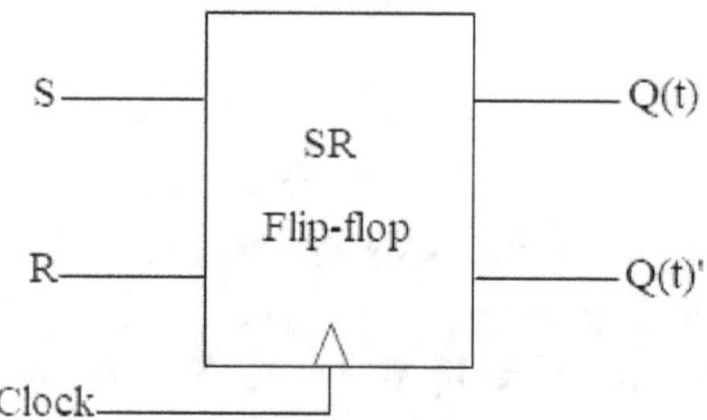

Fig. 12.1 Symbol of SR flip-flop

The SR flip-flop can be constructed using NAND or NOR gate, because both gates are having capability of universal gate. The fig. 12.2. shows SR flip-flop using NAND gates.

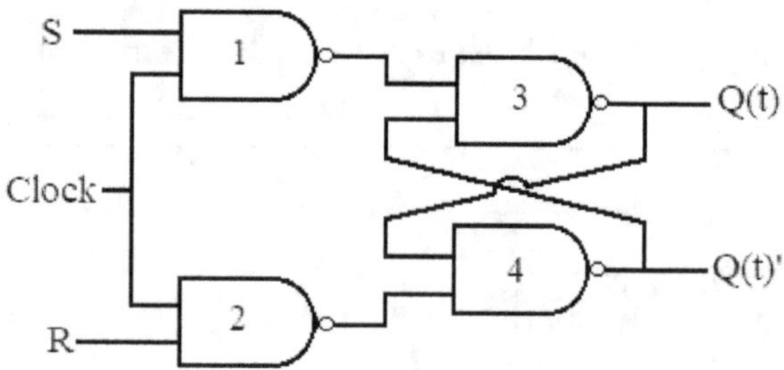

Fig. 12.2 SR flip-flop using NAND gates

The clock pulse plays an important role in operation of flipflop. When clock is absent, the circuit will retain the same state.

Case (I): R = 0 and S = 0

Let us assume Q(t) = 1 and Q(t)' = 0, Clock = 1

The output of gate1 and gate2 will be 1. So, the output of gate3 will be 1 and gate4 output will be 0. It means, same previous value is again at output terminal. This condition is known as 'no change' state.

Case (II): R = 1 and S = 0, Clock = 1

The output of gate1 will be 1 and output of gate2 will be 0. In NAND gate, if any of the inputs is at low level i.e. 0, then output is 1. So, output of gate3 i.e. Q(t)=0 and Q(t)'=1. This is known as reset state.

Case (III): R = 0 and S = 1, Clock = 1

The output of gate1 will be 0 and output of gate2 will be 1. In NAND gate, if any of the inputs is at low level i.e. 0, then output is 1. So, output of gate3 i.e. Q(t)=1 and Q(t)'=0. This is known as set state.

Case (IV): R = 1 and S = 1, Clock = 1

The output of both gate1 and gate2 will be 0. In NAND gate, if any of the inputs is at low level i.e. 0, then output is 1. So, output of gate3 i.e. Q(t)=0 and Q(t)'=0. This is known as prohibited state or indeterminate state, because both the outputs should be complementary in nature. The table 12.1. shows the input and output relationship for SR flip-flop using NAND gate.

Table 12.1 Truth table of SR flip-flop using NAND gate

Inputs			Outputs		State
Clock	S	R	Q(t)	Q(t)'	
0	X	X	NC	NC	No change
1	0	0	NC	NC	No change
1	0	1	0	1	Reset
1	1	0	1	0	Set
1	1	1	1	1	Indeterminate

CHARACTERISTICS TABLE OF SR FLIP-FLOP

A characteristics function or equation of flip-flop is used to explore the value of next state in terms of current state and output. It defines the logical property of flip-flop. The table 12.2. shows characteristics properties of SR flip-flop.

Table 12.2 Characteristics table of SR flip-flop

Clock	S	R	Q(t)	Q(t+1)
1	0	0	0	0
1	0	1	0	0
1	1	0	0	1
1	1	1	0	X
1	0	0	1	1
1	0	1	1	0
1	1	0	1	1
1	1	1	1	X

The characteristics table is filled on the basis of truth table. For example, if S=0, R=0, there is no change. The same is applicable in table 12.3, where the value of previous state will be transferred to next state, without any change. Similarly, all other values are executed. The compiled table for SR flip-flop is shown as in table.12.3.

Table 12.3 Summarised characteristics table of SR flip-flop

S	R	Q(t+1)
0	0	Q(t)
0	1	0
1	0	1
1	1	X (undetermined)

Using K-map, characteristics equation can be explored, as in fig. 12.3.

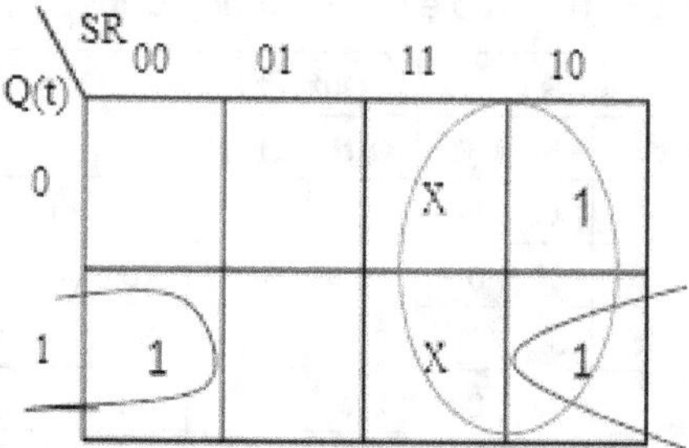

Fig. 12.3 K-map of SR flip-flop characteristics table

$$Q(t + 1) = S + \bar{R}.Q(t)$$

EXCITATION TABLE OF SR FLIP-FLOP

An excitation table of flip-flop shows the necessary input conditions for every possible transition of output. On the basis of the present and next state values, the input conditions are framed. The symbol 'X' signifies don't care condition which means that it does not matter whether the input is 0 or 1. The table 12.4. shows the excitation table of SR flip-flop.

Table 12.4 Excitation table of SR flip-flop

Q(t)	Q(t+1)	S	R
0	0	0	X
0	1	1	0
1	0	0	1
1	1	X	0

The excitation table is deduced from characteristics table. The characteristics table is looked for the values when Q(t)=0 and Q(t+1) =0, the corresponding values for this condition is S = 0, R = 0 and S = 0, R = 1, which can be written as summarised form as: S = 0, R = X. Similarly, the other values of excitation table of SR flip-flops are written.

JK FLIP-FLOP

The problem with SR flip-flop is its undetermined state at values S=1 and R=1. JK flip-flop is refined flipflop, in which undetermined state of SR flip-flop is defined. The symbol of JK flip-flop is shown as in fig. 12.4, where inputs are J and K with corresponding outputs Q(t) and Q(t)'. Basically, the JK flip-flop is combination of SR flip-flop having feedback. The letters J and K are chosen by inventor Jack Kirby.

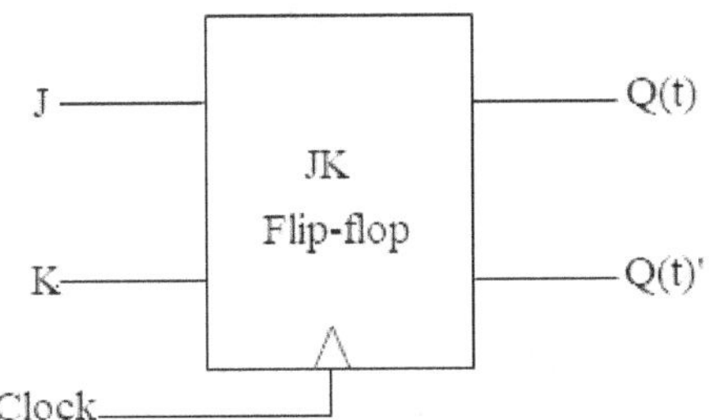

Fig. 12.4 Symbol of JK flip-flop

The JK flip-flop can be constructed using NAND or NOR gate, because both gates are having capability of universal gate. The fig. 12.5. Shows the JK flip-flop using NAND gates.

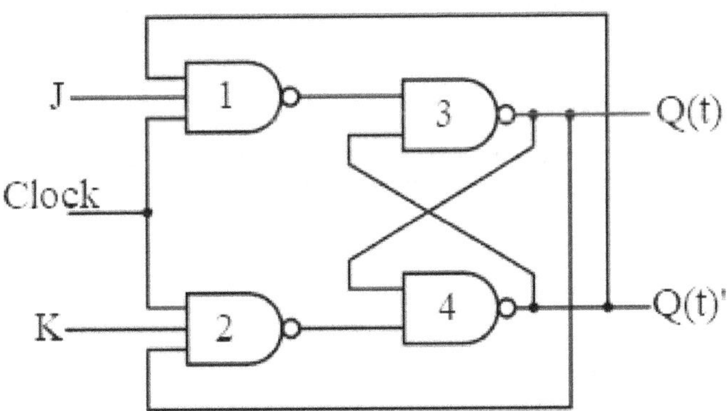

Fig. 12.5 JK flip-flop using NAND gates

The clock pulse plays an important role in operation of flip-flop. When clock is absent, the circuit will retain the same state.

Case (I): J = 0 and K = 0

Let us assume Q(t) = 1 and Q(t)' = 0, Clock = 1

The output of gate1 and gate2 will be 1. So, the output of gate3 will be 1 and gate4 output will be 0. It means, same previous value is again at output terminal. This condition is known as 'no change' state.

Case (II): J = 0 and K = 1

When J = 0 and K = 1, then output of gate1 and gate2 will high. The value of Q(t)' will be fed to gate3 and calculated value will be fed to gate4, this implies Q(t) = 0 i.e. reset condition.

Case (III): J = 1 and K = 0

When J = 1 and K = 0, then output of gate1 will be 0 (assume Q(t)'=1) and gate2 will be 1. The value of Q(t)' will be fed to gate3 and calculated value will be fed to gate4, this implies Q(t) = 1 i.e. set condition.

Case (IV): J=1 and K=1

When J = 1 and K = 0, then output of gate1 will be 0 (assume Q(t)'=1) and gate2 will be 1. The value of Q(t)' will be fed to gate3 and calculated value will be fed to gate4. When the assumed values of Q(t)=0 and Q(t)'=1, then after execution of feedback circuits, the output value of Q(t) =1 and Q(t)'=0. Similarly, when the assumed values of Q(t)=1 and Q(t)'=0, then after execution of feedback circuits, the output value of Q(t) =0 and Q(t)'=1. Hence, toggle condition will exist.

Toggle is an alternate between opposite binary states with application of clock pulse. The input-output relationship of JK flip-flop is shown as in truth table 12.5.

Table 12.5 Truth table of JK flip-flop using NAND gate

Inputs			Outputs		State
Clock	J	K	Q(t)	Q(t)'	
0	X	X	NC	NC	No change
1	0	0	NC	NC	No change
1	0	1	0	1	Reset
1	1	0	1	0	Set
1	1	1	0	1	Toggle

CHARACTERISTICS TABLE OF JK FLIP-FLOP

A characteristics function or equation of flip-flop is used to explore the value of next state in terms of current state and output. It defines the logical property of flip-flop. The table 12.6. shows the characteristics properties of JK flip-flop.

Table 12.6 Characteristics table of JK flip-flop

Clock	J	K	Q(t)	Q(t+1)
1	0	0	0	0
1	0	1	0	0
1	1	0	0	1
1	1	1	0	1
1	0	0	1	1
1	0	1	1	0
1	1	0	1	1
1	1	1	1	0

The characteristics table is filled on the basis of truth table. For example, if J=0, K=0, there is no change. The same is applicable in table 12.7, where the value of previous state will be transferred to next state, without any change. Similarly, all other values are executed.

Table 12.7 Summarised characteristics table of JK flip-flop

J	K	Q(t+1)
0	0	Q(t)
0	1	0
1	0	1
1	1	Toggle

The compiled characteristics table of JK flip-flop is shown as in table 4.8.

Using K-map, characteristics equation can be explored, as in fig. 12.6.

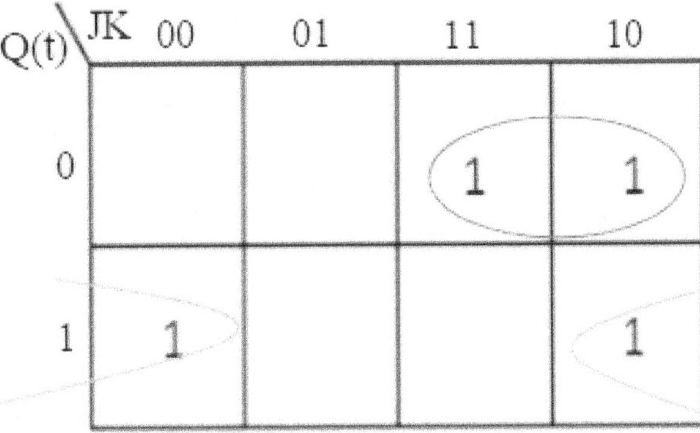

Fig. 12.6 K-map of JK flip-flop characteristics table

$$Q(t+1) = J.Q(t)' + \overline{K}.Q(t)$$

EXCITATION TABLE OF JK FLIP-FLOP

An excitation table of flip-flop shows the necessary input conditions for every possible transition of output. On the basis of the present and next state values, the input conditions are framed. The symbol 'X' signifies don't care condition which means that it does not matter whether the input is 0 or 1. Table 12.8. Shows the excitation properties of JK flip-flop.

Table 12.8 Excitation table of JK flip-flop

Q(t)	Q(t+1)	J	K
0	0	0	X
0	1	1	X
1	0	X	1
1	1	X	0

The excitation table is deduced from characteristics table. The characteristics table is looked for the values when Q(t)=0 and Q(t+1) =0, the corresponding values for this condition is J=0, K=0 and J=0, K=1, which can be written as summarised form as: J=0, K=X. Similarly, the other values of excitation table of JK flip-flops are written.

MASTER SLAVE JK FLIP-FLOP

When J=1 and K=1 in JK flip-flop for long duration, the output Q(t) will toggle as long as clock pulse is high. This condition makes the output of flip-flop unstable. This problem is known as race-around condition in JK flip-flop. This problem can be avoided by ensuring the factor that input clock pulse is high for short duration of time.

The concept of master-slave JK flip-flop is introduced to solve the problem of race around. Here, two JK flip-flops are cascaded with the help of an inverter and feedback channel is also there. One flip-flop act as master flip-flop and another flip-flop acts a slave flip-flop. The clock pulse will determine, which one will be active master flip-flop or slave flip-flop i.e. if clock pulse is high, then Master JK flip-flop will work and slave flip-flop will disable, because its clock pulse will be low.

WORKING OF MASTER-SLAVE FLIP-FLOP

The fig. 12.7. shows the master slave flip-flop. When the clock=1, slave is disabled until clock=1. When clock=0, the information from master flip-flop is transferred to slave flip-flop and output is received. The master flip-flop responds well before the slave flip-flop because it is positive level triggered in comparison with slave flip-flop, which is negative level triggered.

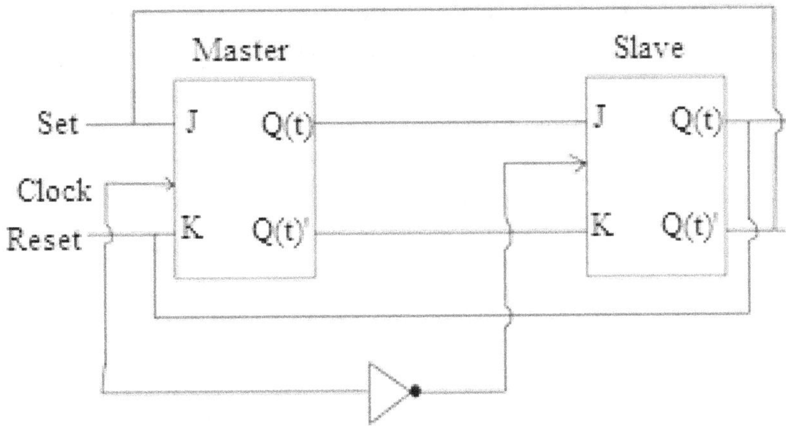

Fig. 12.7 Master-Slave JK flip-flop

Case (I): J=0 and K=0

The flip-flop remains disable and no change in the Q(t).

Case (II): J=0 and K=1

The high Q(t)' output of master flip-flop is fed to K input of slave JK flip-flop and due to clock, the slave is reset. The slave replicates the master flip-flop.

Case (III): J=1 and K=0

The high Q(t) of master flip-flop is fed to J input of slave flip-flop and due to negative transition of clock pulse, the slave flip-flop is active and it copies the master.

Case (IV): J=1 and K=1

It toggles the transition of clock pulse. The master flip-flop toggles on positive clock and slave toggles on negative clock transition.

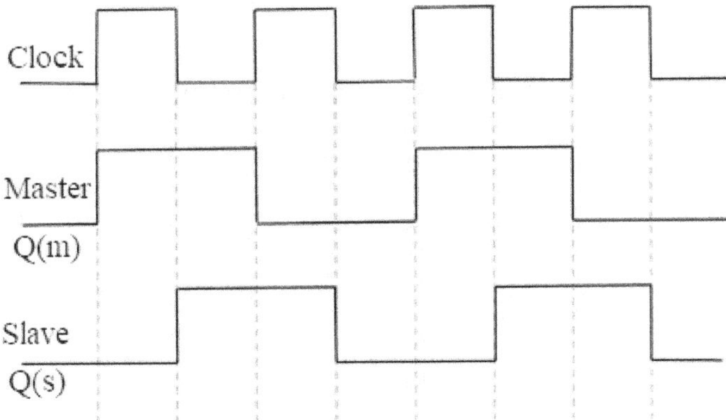

Fig. 12.8 Timing diagram of Master-Slave JK flip-flop

The timing diagram of master-slave flip-flop is shown in fig. 12.8. It is a synchronous device, it passes data with the transition of clock pulse.

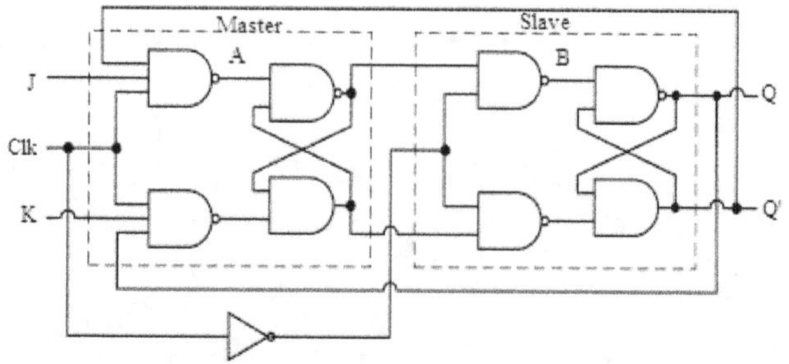

Fig. 12.9 Master-Slave JK flip-flop using NAND gates

The figure 12.9. Shows the circuit diagram of master-slave JK flip-flop using NAND gates.

Table 12.9 Working of Master Slave JK Flip-flop

Clock	J	K	Master	Slave
Positive	1	0	1	Previous State
Negative	1	0	1	1
Positive	0	1	0	Previous State
Negative	0	1	0	0
Positive	1	1	$Q'(t)$	Previous State
Negative	1	1	$Q'(t)$	$Q'(t)$
Positive	0	0	Previous State	Previous State
Negative	0	0	Previous State	Previous State

D FLIP-FLOP

D flip-flop is also known as data flip-flop or transparent flip-flop. In this type of flip-flop, when the clock pulse is active, the output follows its data input. The symbol of D flip-flop is shown as in fig. 12.10.

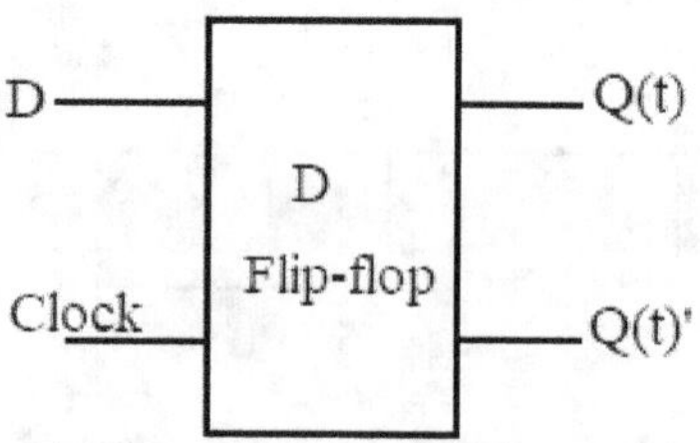

Fig. 12.10 Symbol of D flip-flop

The D flip-flop can be constructed using NAND or NOR gate, because both gates are having capability of universal gate. The fig. 12.11. shows D flip-flop using NAND gates.

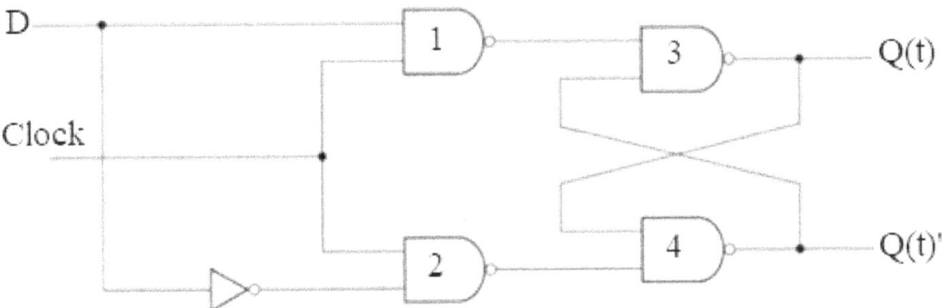

Fig. 12.11 D flip-flop using NAND gates

Case (I): D=0

When D=0 and clock=1, the output of gate1 will be 1 and gate2 will be 0. Assume Q(t)=0 and Q(t)'=1, the value of Q(t)' is fed to gate3 along with output of gate1, which implies Q(t)=0. So, if D=0 then the calculated Q(t) is also 0. It means that same data is transferred.

Case (II): D=1

When D=1 and clock=1, the output of gate1 will be 0 and gate2 will be 1. Assume Q(t)=0 and Q(t)'=1, the value of Q(t)' is fed to gate3 along with output of gate1, which implies Q(t)=1. So, if D=1 then the calculated Q(t) is also 1. It means that same data is transferred. The output does not exist, if the clock pulse is absent. The truth table of D flip-flop is shown in table 12.10.

Table 12.10 Truth table of D flip-flop using NAND gate

Present state			Next state
Clock	D	Q(t)	Q(t+1)
1	0	0	0
1	0	1	0
1	1	0	1
1	1	1	1

CHARACTERISTICS TABLE OF D FLIP-FLOP

A characteristics function or equation of flip-flop is used to explore the value of next state in terms of current state and output. It defines the logical property of flip-flop. The table 12.11. shows the characteristics table of D flip-flop.

Table 12.11 Characteristics table of D flip-flop

Clock	D	Q(t)	Q(t+1)
1	0	0	0
1	1	0	1
1	0	1	0
1	1	1	1

Using K-map, characteristics equation can be explored, as in fig. 12.12.

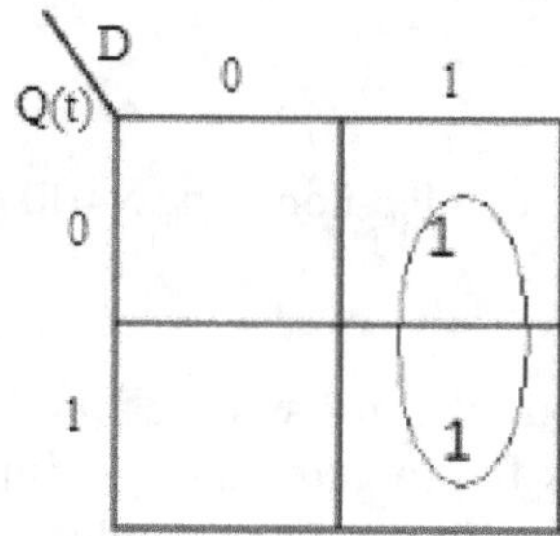

Fig. 12.12 K-map of D flip-flop characteristics table

$$Q(t+1) = D$$

EXCITATION TABLE OF D FLIP-FLOP

An excitation table of flip-flop shows the necessary input conditions for every possible transition of output. On the basis of the present and next state values, the input conditions are framed. The symbol 'X' signifies don't care condition which means that it does not matter whether the input is 0 or 1. The table 12.12. shows excitation properties of D flip-flop.

Table 12.12 Excitation table of D flip-flop

Q(t)	Q(t+1)	D
0	0	0
0	1	1
1	0	0
1	1	1

The excitation table is deduced from characteristics table. The characteristics table is looked for the values when $Q(t) = 0$ and $Q(t+1) = 0$, the corresponding values for this condition is $D = 0$. Similarly, other values of excitation table are filled.

T FLIP-FLOP

T flip-flops are also known as toggle flip-flop. It is the modified form of JK flip-flop. It is designed by connecting J and K terminals together. The symbol of T flip-flop is shown as in fig. 12.13.

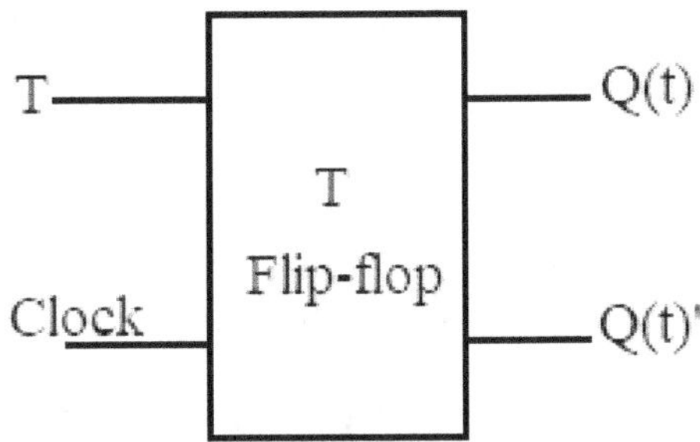

Fig. 12.13 Symbol of T flip-flop

The T flip-flop can be constructed using NAND or NOR gate, because both gates are having capability of universal gate. The fig. 12.14. shows T flip-flop using NAND gate.

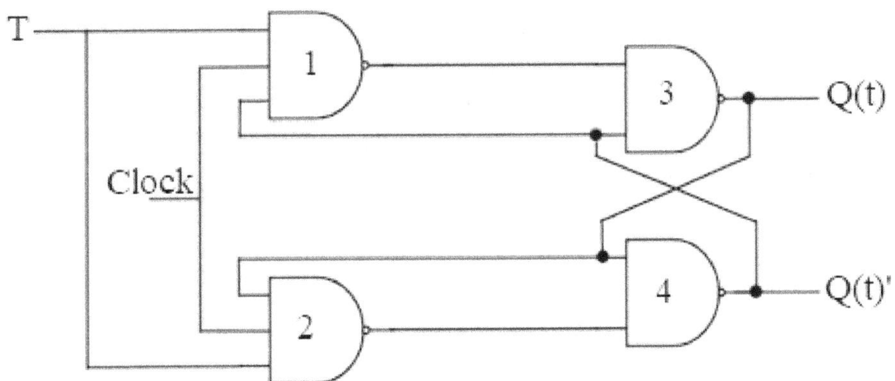

Fig. 12.14 T flip-flop using NAND gates

Case (I): T=0

When T=0 and Clock=1, assume Q(t)=0 and Q(t)'=1. The output of gate1 and gate 2 will be at level 1. The value of Q(t)' is fed to gate3 along with output of gate1, it implies Q(t)=0, hence no change.

Case (II): T=1

When T=1 and Clock=1, assume Q(t)=0 and Q(t)'=1. The output of gate1 will be 0. The value of Q(t)' is fed to gate3 along with output of gate1, it implies Q(t)=1, data is toggled. Similarly, the other case. Toggle means output is complement the present state. The table 12.13. shows input-output relationship of T flip-flop.

Table 12.13 Truth table of T flip-flop using NAND gate

Present state			Next state	State
Clock	T	Q(t)	Q(t+1)	
1	0	0	0	No change
1	0	1	1	
1	1	0	1	Toggle
1	1	1	0	

CHARACTERISTICS TABLE OF T FLIP-FLOP

A characteristics function or equation of flip-flop is used to explore the value of next state in terms of current state and output. It defines the logical property of flip-flop. The table 12.14. shows characteristics properties of T flip-flop

Table 12.14 Characteristics table of T flip-flop

Clock	T	Q(t)	Q(t+1)
1	0	0	0
1	1	0	1
1	0	1	1
1	1	1	0

Using K-map, characteristics equation can be explored, as in fig. 12.15.

Fig. 12.15 K-map of T flip-flop characteristics table

$$Q(t+1) = T.\overline{Q(t)} + \bar{T}.Q(t)$$
$$Q(t+1) = T \oplus Q(t)$$

EXCITATION TABLE OF T FLIP-FLOP

An excitation table of flip-flop shows the necessary input conditions for every possible transition of output. On the basis of the present and next state values, the input conditions are framed. The symbol 'X' signifies don't care condition which means that it does not matter whether the input is 0 or 1. The table 12.15. shows excitation properties of T flip-flop

Table 12.15 Excitation table of T flip-flop

Q(t)	Q(t+1)	T
0	0	0
0	1	1
1	0	1
1	1	0

The excitation table is deduced from characteristics table. The characteristics table is looked for the values when Q(t)=0 and Q(t+1) =0, the corresponding values for this condition is T=0. Similarly, other values of excitation table are filled. For T=0, no change and T=1, toggle is applicable.

EXPERIMENTAL PROCEDURE

1. Check all the ICs using IC tester. After inserting IC carefully in ZIP socket of the tester, the tester should show "PASS".
2. Carefully insert the ICs in breadboard.
3. A square wave generator of frequency 10 Hz (approx.) or multivibrator of longer time period (approx. 1 sec.) is connected to CLK input of each flip-flop.
4. To make SR flip-flop using NAND gates, make the connections as shown in fig. 12.2.
5. Connect LEDs at the output pins of both of the NAND gates.
6. Verify the truth table as given in table 12.1.
7. To make JK flip-flop using NAND gates, make the connections as shown in fig. 12.5.
8. Connect LEDs at the output pins of both of the NAND gates.
9. Verify the truth table as given in table 12.5.

10. To make Master Slave JK flip-flop using NAND gates, make the connections as shown in fig. 12.9.
11. Connect LEDs at the output pins of both of the NAND gates.
12. Verify the truth table as given in table 12.9.
13. To make D flip-flop using NAND gates, make the connections as shown in fig. 12.11.
14. Connect LEDs at the output pins of both of the NAND gates.
15. Verify the truth table as given in table 12.10.
16. To make T flip-flop using NAND gates, make the connections as shown in fig. 12.14.
17. Connect LEDs at the output pins of both of the NAND gates.
18. Verify the truth table as given in table 12.13.

RESULT

(A) The SR flip-flop is designed using NAND gates.
(B) The JK flip-flop is designed using NAND gates.
(C) The Master-Slave JK flip-flop is designed using NAND gates.
(C) The D flip-flop is designed using NAND gates.
(D) The T flip-flop is designed using NAND gates.

PRECAUTIONS

1. Handle the ICs and electronic components carefully.
2. Check the ICs before starting of the experiment. Test the ICs using digital IC tester.
3. The connections should be neat and tight.
4. The LED has one leg long and other leg short. The long leg depicts positive end and short end depicts negative end.
5. Do not press the IC on breadboard until pins are aligned with pours properly.
6. Avoid short circuit and heating of ICs.
7. Ground should be common.
8. For data input +5V must be given for HIGH and 0V for LOW.

VIVA VOCE QUESTIONS

1. What is the need of flip-flop?
2. What is the difference between latch and flip-flop?
3. What is basic difference between clocked SR latch and edge triggered SR flip-flop?

4. Why flip-flops are called sequential circuits?
5. Explain the difference between synchronous and asynchronous inputs?
6. What is race around condition? How it can be overcome?
7. State minimum 4 applications of flip-flops?

Shift Registers

OBJECTIVE

(A)   To design Serial In Serial Out (SISO) shift register using D Flip-flop.

(B)   To design Serial In Parallel Out (SIPO) shift register using D Flip-flop.

(C)   To design Parallel In Serial Out (PISO) shift register using D Flip-flop.

(D)   To design Parallel In Parallel Out (PIPO) shift register using D Flip-flop.

APPARATUS

Breadboard, Power supply (+5V), LEDs, Resistor 220Ω, Connecting wires.

IC REQUIRED

7404, 7474 (D Flip-flop)

INTRODUCTION

Flip-flops are used to store a 1-bit data. For storing a word, group of flip-flops can be used. A register is group of flip-flops. Shift registers are sequential logic circuits that are capable of storing and shifting the data. A n-bit register has group of n flip-flops.

CLASSIFICATION OF SHIFT REGISTERS

Shift registers are used to store and transfer data. They are the combination of flip-flops. A group of flip-flops connected in chain, so that output of one flip-flop becomes the input of another flip-flop. All the flip-flops are driven by a common clock pulse. The basic types of shift registers are shown in fig. 13.1.

1. Serial-In-Serial-Out (SISO) register

2. Serial-In-Parallel-Out (SIPO) register

3. Parallel-In-Serial-Out (PISO) register

4. Parallel-In-Parallel-Out (PIPO) register

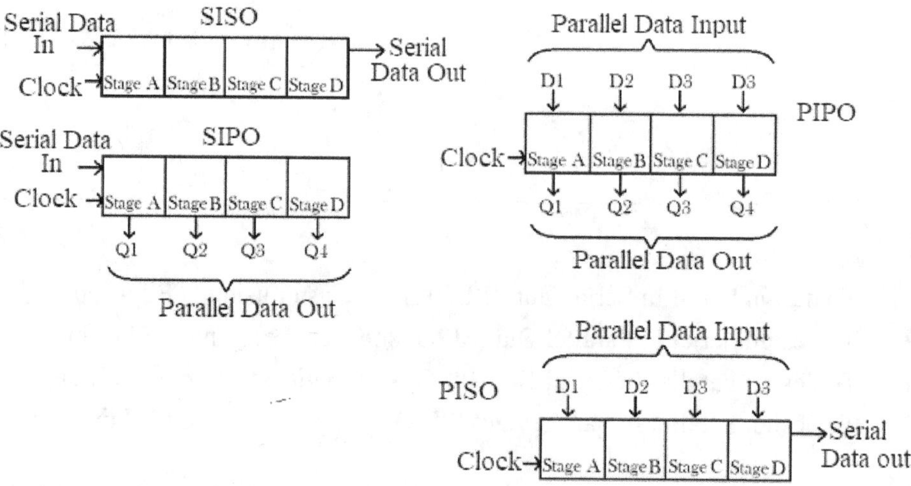

Fig. 13.1 Types of shift registers

SERIAL-IN-SERIAL-OUT (SISO) SHIFT REGISTER

A Serial-In-Serial-Out (SISO) register allows each flip-flop to pass the information from one flip-flop to its adjacent flip-flop serially i.e. one bit by one bit. The movement of data from one end of a shift register to the other at a rate of one bit per clock pulse.

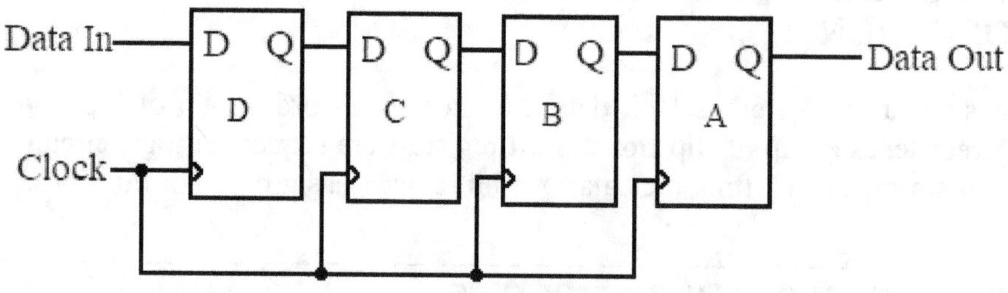

Fig. 13.2 4-Bit shift register (SISO)

The fig. 13.2. represents 4-bit serial-in-serial-out shift register, which consists of four D flip-flops. When the clock pulse is active, data from D flip-flop is shifted to its output Q. Using clear input, all the flip-flops can be vacated initially. Let us take the case of data 1011. On the application of first clock pulse, '1' is shifted to output of first flip-flop i.e. input of second flip-flop. After completion of all the four clock pulses, '1' will reach at

the output of flip-flop4. In this way, the four-bit number is stored in register. For extracting the data out from the shift register, four more clock pulses will be needed.
The function of SISO register is shown as in fig. 13.3.

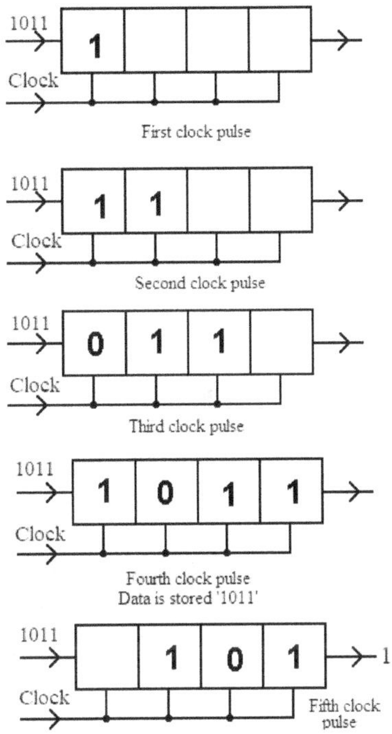

Fig. 13.3 Functioning of SISO register

So, for 4-bit data, storage requires 4 clock pulses and data extracting needs 4 clock pulses.

OBSERVATION TABLE

Take 4-bit data as 1001 and observe the shift operation.(Table 13.1)

Table 13.1 SISO operation of 4-bit binary number 1001

Clock pulse	Output			
	D	C	B	A
0	0	0	0	0
1	1	0	0	0
2	0	1	0	0
3	0	0	1	0
4	1	0	0	1

EXPERIMENTAL PROCEDURE

1. Check all the ICs using IC tester. After inserting IC carefully in ZIP socket of the tester, the tester should show "PASS".
2. Carefully insert the ICs in breadboard.
3. To design SISO using D flip-flop, connect the circuit as shown in fig. 13.2.
4. Enter data 1001 into the register.
5. Verify the truth table as given in table 13.1. In such a way, SISO shift register operation is verified.

RESULT

The Serial -In-Serial-Out Shift register using D Flip-flop is successfully designed.

SERIAL-IN-PARALLEL-OUT (SIPO) SHIFT REGISTER

In Serial-In-Parallel-Out (SIPO) register, the data is fed to the flip-flops serially i.e. one bit by one bit and data is extracted parallelly. The number of flip-flops is equal to number of bits to be stored. The block diagram and function of SIPO register is shown as in fig. 13.4.and fig. 13.5. respectively.

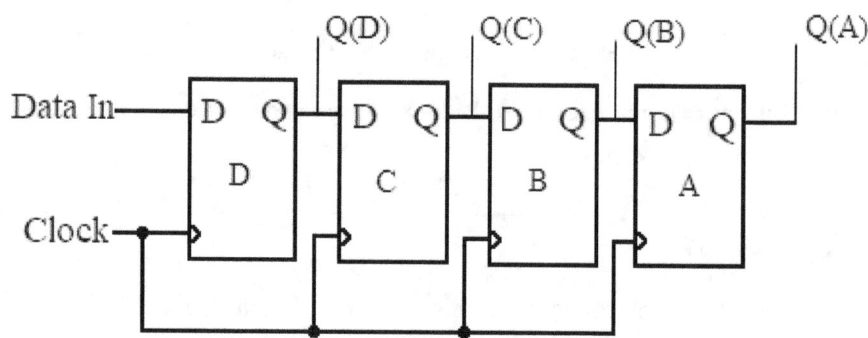

Fig. 13.4 4-bit shift register (SIPO)

Parallel shifting means the movement of data into all the flip-flops of a shift register at the same time i.e. the output from all the flip-flops are extracted simultaneously.

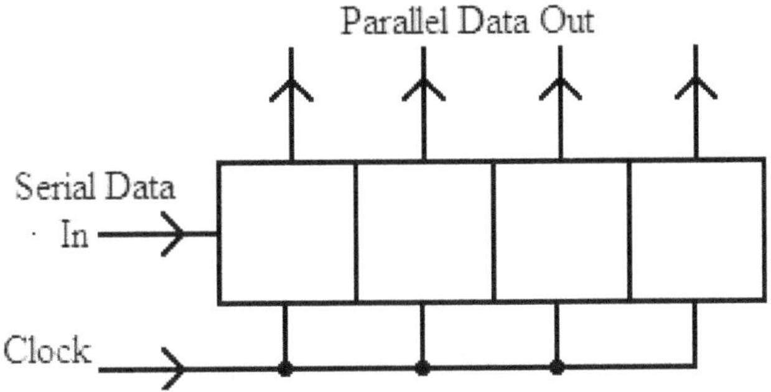

Fig. 13.5 Functioning of SIPO register

So, In SIPO register, for 4-bit data, storage requires 4 clock pulses and data extracting needs a single clock pulse. 4-bit number will be extracted parallelly as a single word.

OBSERVATION TABLE

Take 4-bit data as 1001 and observe the shift operation. (Table 13.2)

Table 13.2 SIPO operation of 4-bit binary number 1001

Clock pulse	Serial Input	Output			
		D	C	B	A
0	0	0	0	0	0
1	1	1	0	0	0
2	0	0	1	0	0
3	0	0	0	1	0
4	1	1	0	0	1

EXPERIMENTAL PROCEDURE

1. Check all the ICs using IC tester. After inserting IC carefully in ZIP socket of the tester, the tester should show "PASS".
2. Carefully insert the ICs in breadboard.
3. To design SIPO using D flip-flop, connect the circuit as shown in fig. 13.4.
4. Enter 4-bit data 1001 into the serial data input with LSB entering first. It will take 4 cycles to shift the 4-bit into the register.
5. Observe the parallel data outputs by connecting LEDs at output D, B, C and A.

6. Verify the truth table as given in table 13.2. In such a way, SIPO shift register operation is verified.

RESULT

The Serial -In-Parallel-Out Shift register using D Flip-flop is successfully designed.

PARALLEL-IN-SERIAL-OUT (PISO) SHIFT REGISTER

In Parallel-In-Serial-Out (PISO) register, the data is fed to all the flip-flops simultaneously. While extracting the data from the flip-flops, it is sequential process i.e. one bit by one bit.

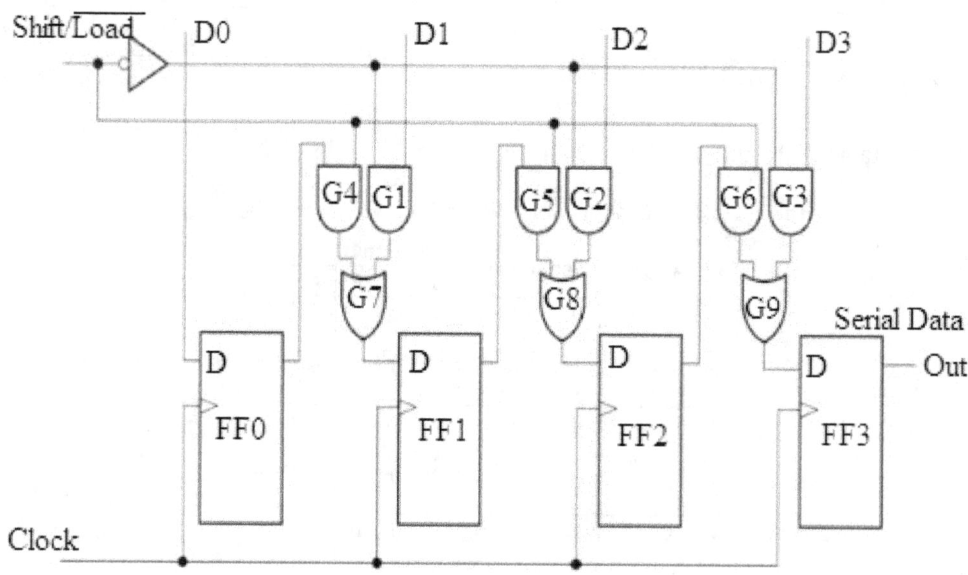

Fig. 13.6 4-bit shift register (PISO)

The block diagram and function of PISO register are shown as in fig. 13.6 and fig. 13.7. respectively. In 4-bit PISO register, four flip-flops are used. Data D0 D1 D2 D3 is entered parallelly. $Shift/\overline{Load}$ is the control unit.

When $Shift/\overline{Load}$ is low i.e. '0', then logic gates G1, G2 and G3 are enabled, which allow input data to be applied to D input of its respective flip-flop. When the clock pulse is applied, the flip-flop with value D=1 will set and D=0 will reset.

When $Shift/\overline{Load}$ is high i.e. '1', then logic gates G1, G2 and G3 are disabled and logic gates G4, G5 and G6 are enabled. This allows the shifting operation.

The gates G7, G8 and G9 at the D inputs of the flip-flops allow either parallel data entry or shift operation, depending upon the gates that are enabled by $Shift/\overline{Load}$ control paramete.

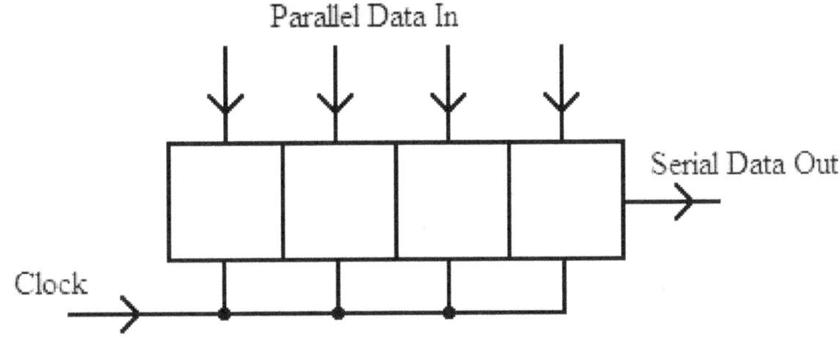

Fig. 13.7 Functioning of PISO register

So, In PISO register, for 4-bit data, storage requires single clock pulse for storing and 4 clock pulses for shifting operation.

OBSERVATION TABLE

Take 4-bit data as 1001. Shift/$\overline{Load}$ is high and clock pulse is enabled. (Table 13.3)

Table 13.3 PISO operation of 4-bit binary number 1001

Clock pulse	Output				Serial Data Output
	D	**C**	**B**	**A**	
1	1	0	0	1	1
2	0	1	0	0	0
3	0	0	1	0	0
4	1	0	0	1	1

EXPERIMENTAL PROCEDURE

1. Check all the ICs using IC tester. After inserting IC carefully in ZIP socket of the tester, the tester should show "PASS".
2. Carefully insert the ICs in breadboard.
3. To design PISO using D flip-flop, connect the circuit as shown in fig. 13.6.
4. Take Shift/$\overline{Load}$ input as low.
5. Now enter 4-bit input data into flip-flops.
6. On the application of clock pulse, 4-bit data gets loaded into 4 flip-flops.

7. Now for performing serial shifting operation, take Shift/$\overline{Load}$ input as '1'.

8. Apply clock pulse, verify the serial data output as per the observation table 13.3.

RESULT

The Serial -In-Parallel-Out Shift register using D Flip-flop is successfully designed.

PARALLEL-IN-PARALLEL-OUT (PIPO) SHIFT REGISTER

Parallel-in-parallel-out PIPO register is a type of shift registers in which data entering and data extracting are parallel processes. The PIPO can be preset to any value by directly loading a binary number into its internal flip-flops and also extract the output simultaneously. The block diagram and function of PIPO register are shown as in fig. 13.8 and fig. 13.9 respectively.

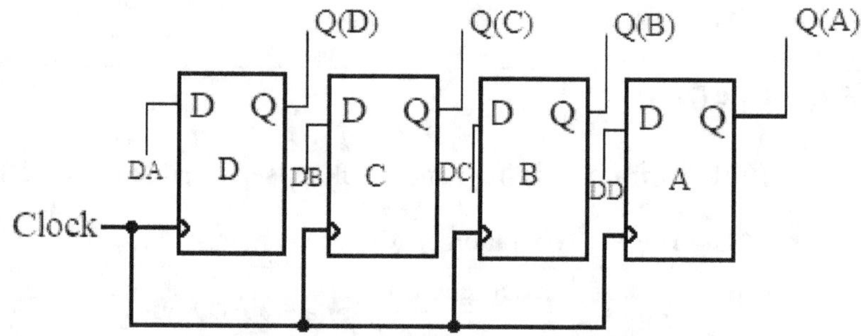

Fig. 13.8 4-bit shift register (PIPO)

Parallel shifting means the movement of data into all the flip-flops of a shift register at the same time i.e. the output from all the flip-flops are extracted simultaneously.

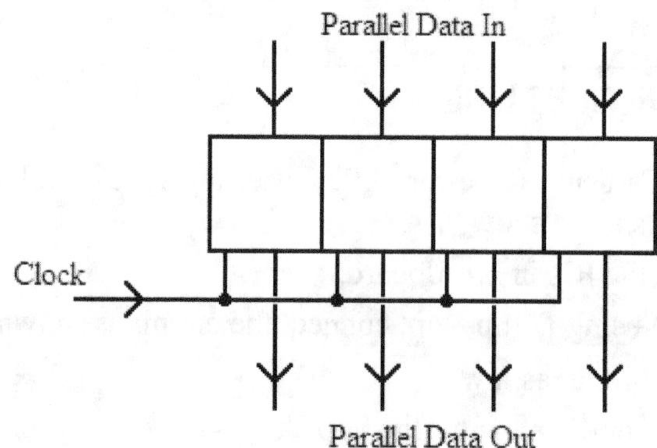

Fig. 13.9 Functioning of PIPO register

So, in 4-bit PIPO shift register operation, one clock pulse is required for data entering and one clock pulse is required for extraction of data. In total, two clock pulses are consumed by PIPO shift register for data store and shift purpose.

OBSERVATION TABLE

Take 4-bit data as 1001 and observe the shift operation. (Table 13.4)

Table 13.4 PIPO operation of 4-bit binary number 1001

Clock pulse	Output			
	D	C	B	A
1	1	0	0	1
2	1	0	0	1

EXPERIMENTAL PROCEDURE

1. Check all the ICs using IC tester. After inserting IC carefully in ZIP socket of the tester, the tester should show "PASS".
2. Carefully insert the ICs in breadboard.
3. To design PIPO using D flip-flop, connect the circuit as shown in fig. 13.8.
4. Enter data 1001 into the register.
5. Verify the truth table as given in table 13.4. In such a way, PIPO shift register operation is verified.

RESULT

The Parallel-In-Parallel-Out Shift register using D Flip-flop is successfully designed.

PRECAUTIONS

1. Handle the ICs and electronic components carefully.
2. Check the ICs before starting of the experiment. Test the ICs using digital IC tester.
3. The connections should be neat and tight.
4. The LED has one leg long and other leg short. The long leg depicts positive end and short end depicts negative end.
5. Do not press the IC on breadboard until pins are aligned with pours properly.
6. Avoid short circuit and heating of ICs.

7. Ground should be common.
8. For data input +5V must be given for HIGH and 0V for LOW.

VIVA VOCE QUESTIONS

1. What is the need and applications of shift register?
2. What is the difference between register and flip-flop?
3. What is the significance of IC 74165 and IC74164?
4. What do you mean by universal shift register? Why it is known as universal? What is its advantages?
5. What is the pin diagram of universal shift register IC 74194?
6. How many bits a register can store?
7. Why shifting is not possible in PIPO shift register?

Experiment 14

Shift Register Counter

OBJECTIVE

(A)   To design Ring Counter using D Flip-flop.

(B)   To design Johnson Counter using D Flip-flop.

APPARATUS

Breadboard, Power supply (+5V), LEDs, Resistor 220Ω, Connecting wires.

IC REQUIRED

7404, 7474 (D Flip-flop)

INTRODUCTION

Shift registers counters are designed using SISO shift registers with feedback from output of last flip-flop to input of first flip-flop. As they exhibit specified sequence of states, therefore such devices are known as counters. Commonly used shift register counter is:

1. Ring counter   2. Johnson counter

4.7.2.1 RING COUNTER

In ring counter, number of flip-flops depend on the number of data bits. The arrangement of flip-flop array is in form of ring, so it is known as ring counter. The output of first flip-flop is connected to input of second flip-flop and so on. The output of last flip-flop becomes the input of first flip-flop by feedback manner. The fig. 14.1. shows 4-bit ring-counter using D flip-flop.

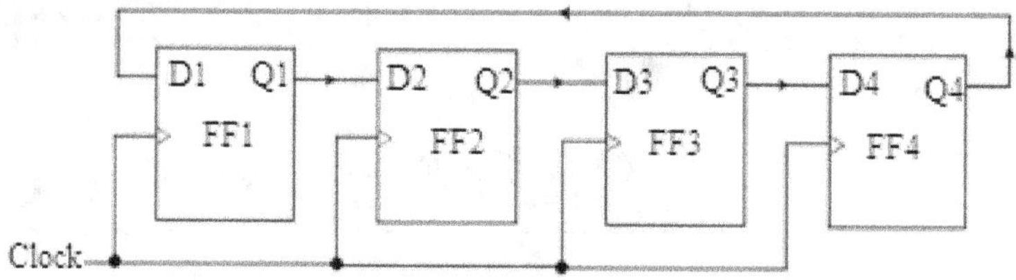

Fig. 14.1 4-bit ring counter

The fig. 14.1. shows 4-bit ring counter, designed using D flip-flops. It is a circular shift register, where one flip-flop is set and all other flip-flops are clear. One bit is shifted within the flip-flops to produce sequence of timing signals. Let us take a case of 4-bit data 1100 to be shifted and circulated back with respect to clock signals. The table 14.1. shows the sequence state table of ring counter.

Table 14.1 Sequence table of ring counter

Clock pulse	Q1	Q2	Q3	Q4
-	1	1	0	0
1	0	1	1	0
2	0	0	1	1
3	1	0	0	1
4	1	1	0	0
5	0	1	1	0
6	0	0	1	1
7	1	0	0	1

The state diagram of 4-bit ring counter is shown as in fig. 14.2.

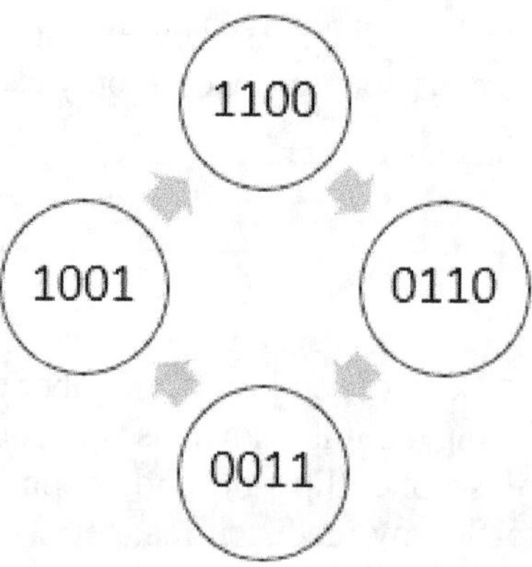

Fig. 14.2 State diagram of 4-bit ring counter

EXPERIMENTAL PROCEDURE

1. Check all the ICs using IC tester. After inserting IC carefully in ZIP socket of the tester, the tester should show "PASS".
2. Carefully insert the ICs in breadboard.
3. To design Ring counter using D flip-flop, connect the circuit as shown in fig. 14.1.
4. Enter 4-bit data 1100 into the shift register counter.
5. Verify the truth table as given in table 14.1. In such a way, Ring counter operation is verified.

RESULT

The Ring counter using D Flip-flop is successfully designed.

JOHNSON COUNTER

The Johnson counter is also known as twisted ring counter or switch-rail counter. The output of first flip-flop is connected to input of second flip-flop and so on. The complemented output of last flip-flop becomes the input of first flip-flop by feedback manner. Therefore, it is named as twisted ring counter. The number of states can be doubled if the shift register is connected as a switch-trail ring counter.

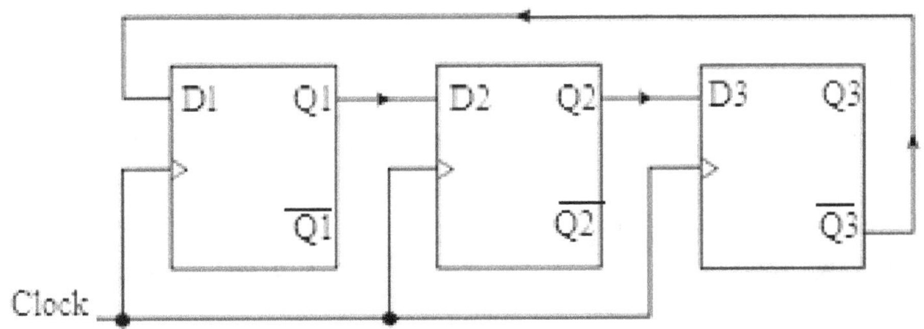

Fig. 14.3 3-bit Johnson counter

In fig. 14.3. 3-bit Johnson counter using D flip-flop is shown. The number of flip-flops is equal to number of data bits. Let us reset all the flip-flops initially. The output Q1 is attached with D2, Q2 is connected with D3 and complemented output Q3 becomes the input for D1. The sequence state table of Johnson counter is shown in table 14.2.

Table 14.2 Sequence table of Johnson counter

Clock pulse	Q1	Q2	Q3
-	0	0	0
1	1	0	0
2	1	1	0
3	1	1	1
4	0	1	1
5	0	0	1
6	0	0	0
7	1	0	0

The state diagram of 3-bit Johnson counter is shown as in fig. 14.4.

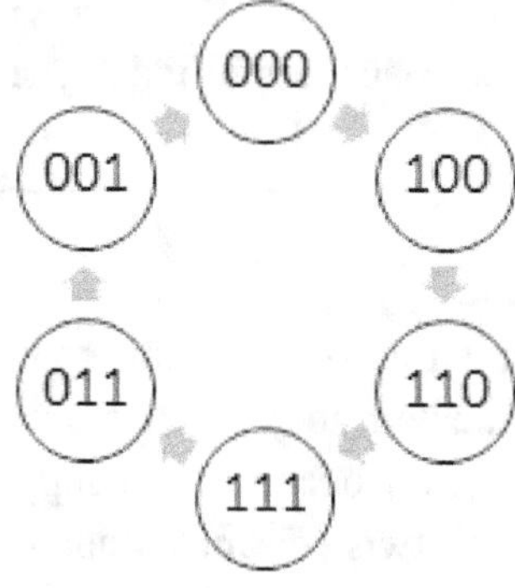

Fig. 14.4 State diagram of 3-bit Johnson counter

EXPERIMENTAL PROCEDURE

1. Check all the ICs using IC tester. After inserting IC carefully in ZIP socket of the tester, the tester should show "PASS".
2. Carefully insert the ICs in breadboard.
3. To design Johnson counter using D flip-flop, connect the circuit as shown in fig. 14.3.
4. Enter 3-bit data 100 into the shift register counter.
5. Verify the truth table as given in table 14.2. In such a way, Johnson counter operation is verified.

RESULT

The Ring counter using D Flip-flop is successfully designed.

PRECAUTIONS

1. Handle the ICs and electronic components carefully.
2. Check the ICs before starting of the experiment. Test the ICs using digital IC tester.
3. The connections should be neat and tight.
4. The LED has one leg long and other leg short. The long leg depicts positive end and short end depicts negative end.
5. Do not press the IC on breadboard until pins are aligned with pours properly.
6. Avoid short circuit and heating of ICs.
7. Ground should be common.
8. For data input +5V must be given for HIGH and 0V for LOW.

VIVA VOCE QUESTIONS

1. What is the significance of shift register counter?
2. What is Ring counter and Johnson counter?
3. What is the application of Ring counter and Johnson counter?
4. What do you mean by state diagram and state table?
5. What is the pin diagram of IC7474?
6. What do you understand by short circuit?
7. Why Johnson counter is also known as twisted ring counter?

Asynchronous & Synchronous Counter

OBJECTIVE

(A) To design MOD-4/2-bit Asynchronous counter/Ripple counter using T Flip-flop.

(B) To design MOD-4/2-bit Synchronous counter using JK Flip-flop.

(C) To design 3-bit synchronous up/down counter using T flip-flop

APPARATUS

Breadboard, Power supply (+5V), LEDs, Resistor 220Ω, Connecting wires.

IC REQUIRED

74107 (T Flip-flop), 7476 (JK Flip-flop)

INTRODUCTION

A counter is a sequential circuit that is capable of counting the number of clock pulses arriving at its clock input. A counter is a set of flip-flops whose states change in response to pulses applied at the input of the counter. The flip-flops are interconnected such that their combined state at any time is the binary equivalent of the total number of pulses that have occurred up to that time. One of the applications of counter is frequency divider.

A n-bit counter consists of n flip-flops and can count from 0 to 2^{n-1}. It is also called Mod-N counter, where N is the number of states.

CLASSIFICATION OF COUNTERS

Basically, counters are divided into two categories:

1. Asynchronous counter

2. Synchronous counter

Asynchronous counters are also known as ripple counter, serial counter or non-synchronous counter. In this type of counter, each flip-flop is triggered by previous flip-flop and the first flip-flop is triggered by external clock pulse. In synchronous counter, all the flip-flops are triggered simultaneously by common clock pulse.

ASYNCHRONOUS COUNTERS

In asynchronous counters, the number of flip-flops depends on the number of states. The number of output states of the counter is called Modules (MOD). If we have three flip-flops, the maximum number of states will be eight. We can name that counter which has eight output states, as MOD-8 counter.

MOD-4/2-BIT ASYNCHRONOUS COUNTER/RIPPLE COUNTER

In 2-bit asynchronous counter, two flip-flops are used. As the number of flip-flops are two, the possible states are $2^2=4$. Depending on the number of states, it is known as MOD 4 counter.

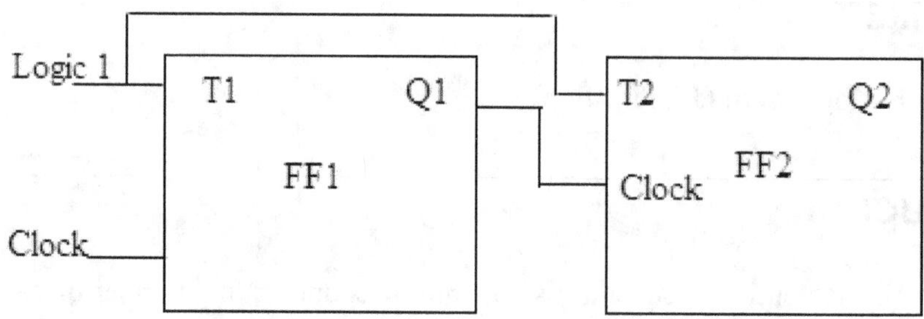

Fig. 15.1 Asynchronous MOD-4 counter using T flip-flop

The fig. 15.1. shows Asynchronous MOD-4 counter using T flip-flop. Initially, both flip-flops are cleared i.e. Q2Q1=00.

As soon as negative edge clock pulse is applied, FF1 will toggle and Q1=1. This output Q1 is connected to second flip-flop as clock. With transition, it becomes positive edge clock, so there will be no change in Q2. In such a manner, Q2Q1=01

When second clock pulse is applied, the FF1 will toggle, Q1=0 which is connected to FF2. The change will act as negative edge clock pulse, FF2 will toggle and Q2=1. In such a manner, Q2Q1=10.

When 3rd clock pulse is applied, FF1 will toggle and Q1=0. This output Q1 is connected to second flip-flop as clock. With transition, it becomes positive edge clock, so there will be no change in Q2. In such a manner, Q2Q1=11

When 4th clock pulse is applied, FF1 toggles again and Q1=1. The change will act as negative edge clock pulse, FF2 will toggle and Q2=0. In such a manner, Q2Q1=00.

Table 15.1 Truth table of MOD-4 Asynchronous counter

Clock pulse	Counter outputs		State	Decimal equivalent
	Q2	Q1		
-	0	0	-	0
1st	0	1	1	1
2nd	1	0	2	2
3rd	1	1	3	3
4th	0	0	4	0

The input-output relationship of MOD-4 asynchronous counter is shown as in table 15.1.

ADVANTAGES & DISADVANTAGES OF ASYNCHRONOUS COUNTERS

Advantages

1. Simple design
2. Number of components are less
3. Working is understandable easily

Disadvantages

1. As flip-flops are not clocked simultaneously, so speed is slow.
2. For truncated sequence, these counters are forced to recycle before going through all of its normal states.
3. Propagation delay is high. It limits the rate at which counter can be clocked.

EXPERIMENTAL PROCEDURE

1. Check all the ICs using IC tester. After inserting IC carefully in ZIP socket of the tester, the tester should show "PASS".
2. Carefully insert the ICs in breadboard.

3. To design 2-bit asynchronous counter using T flip-flop, connect the circuit as shown in fig. 15.1.

4. Verify the truth table as given in table 15.1. In such a way, 2-bit asynchronous counter operation is verified.

RESULT

The 2-bit asynchronous counter using T flip-flop is successfully designed.

SYNCHRONOUS COUNTER

If the clock pulse is applied to all flip-flops simultaneously, then it is known as synchronous counter i.e. when same clock pulse triggers all the flip-flops, it is known as synchronous counter. In synchronous counters, the number of flip-flops depends on the number of states. The number of output states of the counter is called Modules (MOD). If we have two flip-flops, the maximum number of states will be four. We can name that counter which has four output states, as MOD-4 counter.

MOD-4/2-BIT SYNCHRONOUS COUNTER USING JK FLIP-FLOP

In 2-bit synchronous counter, two flip-flops are used. As the number of flip-flops are two, the possible states are $2^2=4$. Depending on the number of states, it is known as MOD 4 counter.

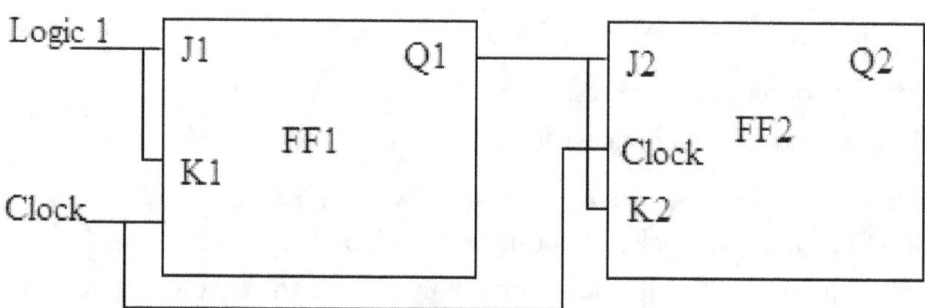

Fig. 15.2 Synchronous MOD-4 counter using JK flip-flop

Fig. 15.2. represents synchronous MOD-4/2-bit synchronous counter using JK flip-flop. The input J and K are connected together and tied at logic 1. Same clock pulse is given to FF1 and FF2.

Initially, both flip-flops are cleared i.e. Q2Q1=00.

As soon as negative edge clock pulse is applied, FF1 will toggle, because J1=K1=1, whereas Q2 will remain unchanged because J2=K2=0 and Q1=1, at the end of pulse. So, after first clock pulse Q2Q1=01

When second clock pulse is applied, the both FF1 and FF2 will toggle. Thus, after second clock pulse, Q2Q1=10

When 3rd clock pulse is applied, FF1 will toggle and Q1=1. There will be no change in Q2. In such a manner, Q2Q1=11

When 4th clock pulse is applied, FF1 and FF2 will toggle again as J1=K1=J2=K2=1. This results in Q2Q1=00. So, the counter is recycled back to its original state.

Table 15.2 Truth table of MOD-4 Synchronous counter

Clock pulse	Counter outputs		State	Decimal equivalent
	Q2	Q1		
-	0	0	-	0
1st	0	1	1	1
2nd	1	0	2	2
3rd	1	1	3	3
4th	0	0	4	0

The table 15.2. shows the input-output relationship of MOD-4 synchronous counter

EXPERIMENTAL PROCEDURE

1. Check all the ICs using IC tester. After inserting IC carefully in ZIP socket of the tester, the tester should show "PASS".
2. Carefully insert the ICs in breadboard.
3. To design 2-bit synchronous counter using JK flip-flop, connect the circuit as shown in fig. 15.2.
4. Verify the truth table as given in table 15.2. In such a way, 2-bit synchronous counter operation is verified.

RESULT

The 2-bit synchronous counter using JK flip-flop is successfully designed.

3-BIT UP/DOWN COUNTER USING T FLIP-FLOP

An up/down counter is also known as bidirectional counter. As the both operations up and down have been performed in a single circuit, so its operation is controlled by $up/\overline{down}$. When this signal is low, decremented/down operation will be performed and when this signal is at high level, up/incremented operation will be performed.

STATE DIAGRAM

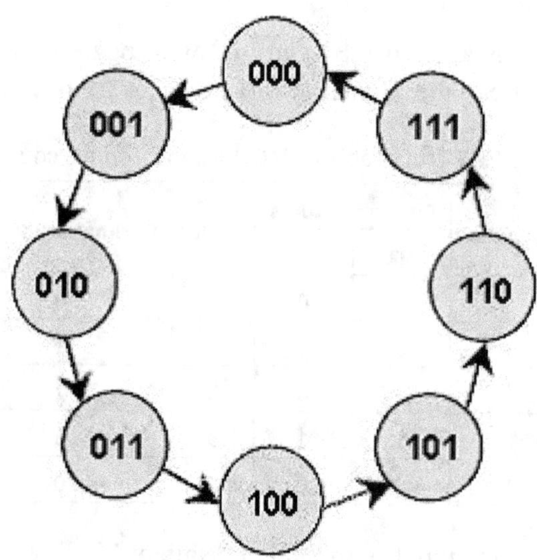

Fig. 15.3 State diagram of 3-bit counter

STATE TABLE

The state diagram and state table of 3-bit up/down counter is shown as in fig. 15.3. and table 15.3. respectively.

Table 15.3 State table 3-bit Up/down counter

Input	Present state			Next state			Flip-flop inputs		
$up/\overline{down}$	Q_C	Q_B	Q_A	Q_{C+1}	Q_{B+1}	Q_{A+1}	T_C	T_B	T_A
0	0	0	0	1	1	1	1	1	1
0	0	0	1	0	0	0	0	0	1
0	0	1	0	0	0	1	0	1	1
0	0	1	1	0	1	0	0	0	1
0	1	0	0	0	1	1	1	1	1
0	1	0	1	1	0	0	0	0	1
0	1	1	0	1	0	1	0	1	1
0	1	1	1	1	1	0	0	0	1

Table 15.3 *Contd...*

Input	Present state			Next state			Flip-flop inputs		
$up/\overline{down}$	Q_C	Q_B	Q_A	Q_{C+1}	Q_{B+1}	Q_{A+1}	T_C	T_B	T_A
1	0	0	0	0	0	1	0	1	1
1	0	0	1	0	1	0	0	0	1
1	0	1	0	0	1	1	0	1	1
1	0	1	1	1	0	0	1	0	1
1	1	0	0	1	0	1	0	1	1
1	1	0	1	1	1	0	0	0	1
1	1	1	0	1	1	1	0	1	1
1	1	1	1	0	0	0	1	0	1

The logical expressions can be found using K-map

$U_D Q_C$ \\ $Q_B Q_A$	00	01	11	10
00	1			
01	1			
11			1	
10			1	

$$TC = \overline{UD}.\overline{QB}.\overline{QA} + UD.QB.QA$$

$U_D Q_C$ \\ $Q_B Q_A$	00	01	11	10
00	1			1
01	1			1
11		1	1	
10		1	1	

$$TB = \overline{UD}.\overline{QA} + UD.QA$$

$U_D Q_C$ \\ $Q_B Q_A$	00	01	11	10
00	1	1	1	1
01	1	1	1	1
11	1	1	1	1
10	1	1	1	1

$$TC = 1$$

After extracting the logical expression, the 3-bit up/down synchronous counter is designed using T flip-flop as in fig. 15.4.

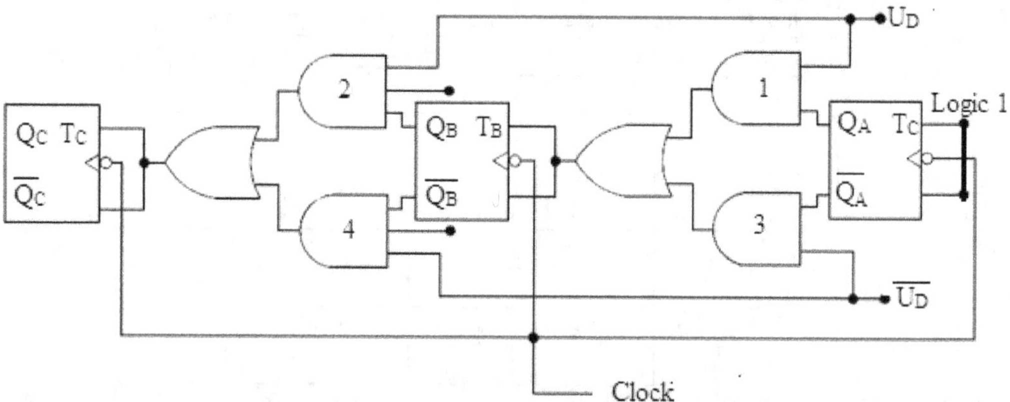

Fig. 15.4 3-bit synchronous up/down counter using T flip-flop

Fig. 15.4. shows the design of 3-bit up/down counter which performs increment as well as decrement operation, using the same logic circuit.

EXPERIMENTAL PROCEDURE

1. Check all the ICs using IC tester. After inserting IC carefully in ZIP socket of the tester, the tester should show "PASS".
2. Carefully insert the ICs in breadboard.
3. To design 3-bit Up/Down counter using T flip-flop, connect the circuit as shown in fig. 15.4.
4. Verify the truth table as given in table 15.3. In such a way, 3-bit Up/Down counter operation is verified.

RESULT

The 3-bit Up/Down counter using T flip-flop is successfully designed.

PRECAUTIONS

1. Handle the ICs and electronic components carefully.
2. Check the ICs before starting of the experiment. Test the ICs using digital IC tester.
3. The connections should be neat and tight.
4. The LED has one leg long and other leg short. The long leg depicts positive end and short end depicts negative end.
5. Do not press the IC on breadboard until pins are aligned with pours properly.

6. Avoid short circuit and heating of ICs.
7. Ground should be common.
8. For data input +5V must be given for HIGH and 0V for LOW.

VIVA VOCE QUESTIONS

1. What is the significance of synchronous and asynchronous counter?
2. What is ripple counter? Give its applications?
3. What is the significance of clock?
4. What do you mean by excitation table and state table?
5. What is the pin diagram of IC7476?
6. What do you understand by CLEAR and PRESET?
7. Can a single counter act as UP counter as well as DOWN counter?

Decade Counter on 7-Segment Display

OBJECTIVE

To design decade counter and visualize its output on 7-segment display.

APPARATUS

Breadboard, Power supply (+5V), LEDs, Resistor 220Ω, Connecting wires, Function generator, seven segment display.

IC REQUIRED

7493 (Binary counter), 7447 (7-segment decoder)

INTRODUCTION

A counter is a sequential circuit that is capable of counting the number of clock pulses arriving at its clock input. A counter is a set of flip-flops whose states change in response to pulses applied at the input of the counter. The flip-flops are interconnected such that their combined state at any time is the binary equivalent of the total number of pulses that have occurred up to that time. One of the applications of counter is frequency divider.

A n-bit counter consists of n flip-flops and can count from 0 to 2^{n-1}. It is also called Mod-N counter, where N is the number of states.

The design of a synchronous sequential circuits starts from a set of specifications and culminates in a logic diagram or a list of Boolean functions from which a logic diagram can be obtained. In contrast to a combinational logic, which is fully specified by a truth table, a sequential circuit requires a state table for its specification. Steps for the design of sequential circuits are shown in Figure 16.1.

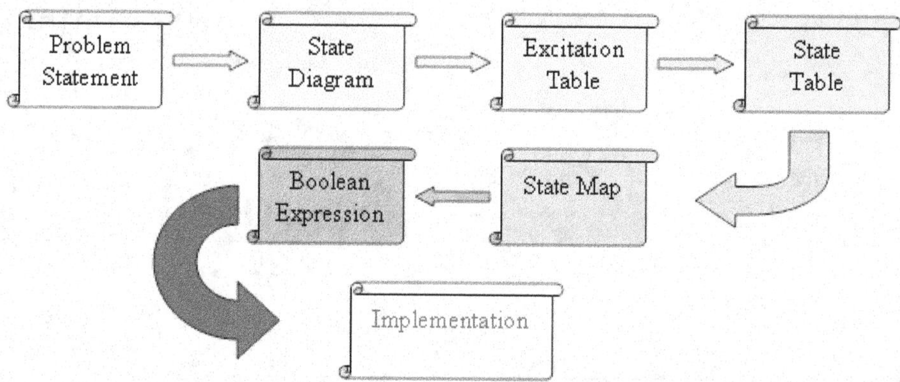

Fig. 16.1 Steps to design sequential circuits.

A synchronous sequential circuit is made of flip-flops and combinational gates. The design of the circuit consists of choosing the flip-flops and then finding the combinational structure which, together with the flip-flops, produces a circuit that fulfils the required specifications. The number of flip-flops is determined from the number of states needed in the circuit.

DECADE COUNTER

A decade counter counts in a sequence of ten and then returns back to zero after the count of nine. So, to count a binary value of nine, the counter must have at least four flip-flops within its chain to represent each decimal digit as shown in fig. 16.1(a). The decade counter has four flip-flops and 16 potential states, of which only 10 are used. The decade counter is also known as MOD 10 counter, that counts from 0 to 9.

STATE DIAGRAM

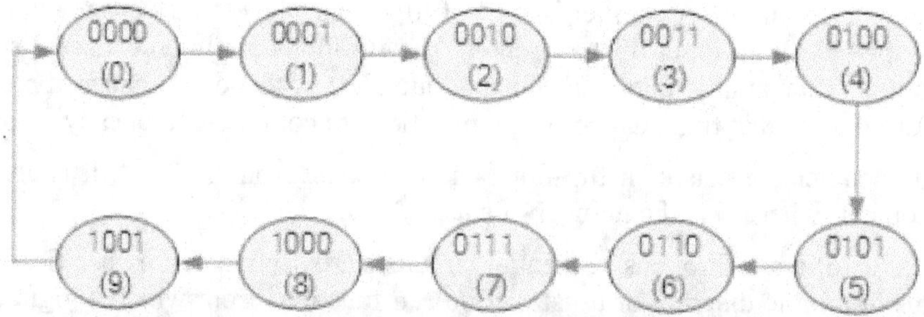

Fig. 16.1(a) State diagram of decade counter

It is called a BCD counter because its ten-state sequence is that of a BCD code and does not have a regular pattern, unlike a straight binary counter.

STATE TABLE

Table 16.1 State table of decade counter

Present state				Next state			
Q_D	Q_C	Q_B	Q_A	Q_{D+1}	Q_{C+1}	Q_{B+1}	Q_{A+1}
0	0	0	0	0	0	0	1
0	0	0	1	0	0	1	0
0	0	1	0	0	0	1	1
0	0	1	1	0	1	0	0
0	1	0	0	0	1	0	1
0	1	0	1	0	1	1	0
0	1	1	0	0	1	1	1
0	1	1	1	1	0	0	0
1	0	0	0	1	0	0	1
1	0	0	1	0	0	0	0

IC7493 PIN DIAGRAM

The IC 7493 is used for decade counter. The fig. 16.2. describes its pin diagram.

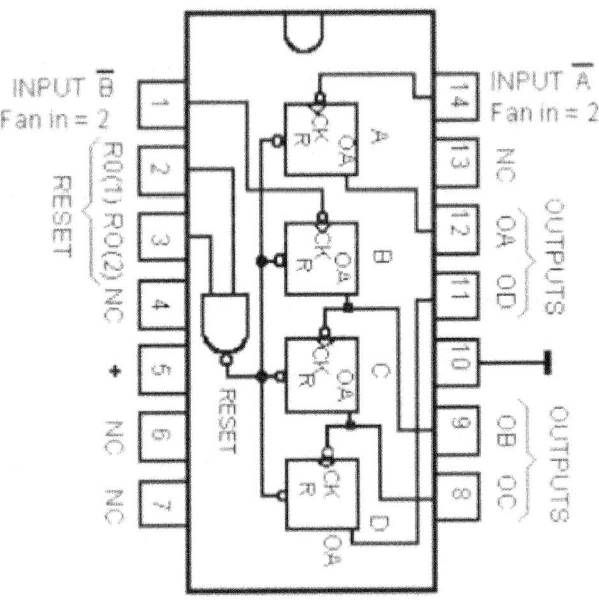

Fig. 16.2 Pin diagram of IC7493

DECADE COUNTER INTERFACING WITH 7447

The decade counter is interfaced with seven segment display using 7447 decoder. The interfacing diagram is shown as in fig. 16.3.

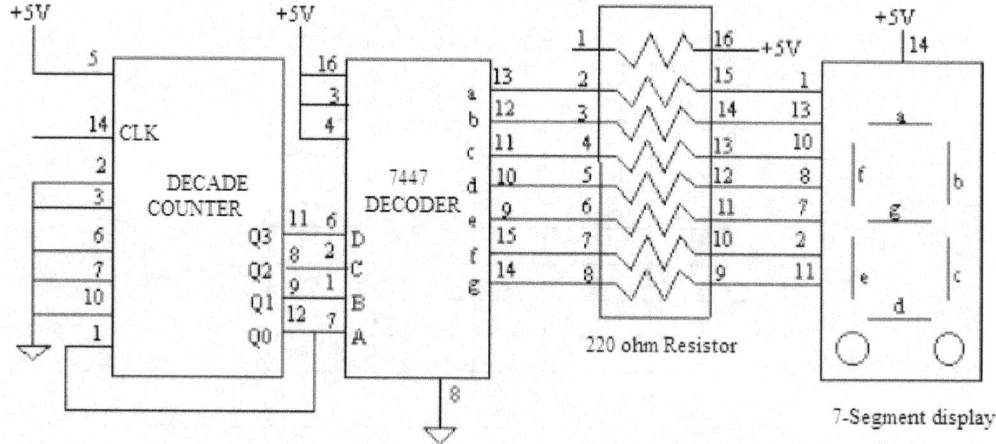

Fig. 16.3 Output of Decade counter on 7-segment display

EXPERIMENTAL PROCEDURE

1. Check all the ICs using IC tester. After inserting IC carefully in ZIP socket of the tester, the tester should show "PASS".
2. Carefully insert the ICs in breadboard.
3. To design decade counter and visualize its output on seven segment display, connect the circuit as shown in fig. 16.3.
4. Apply clock pulse of frequency 1Hz.
5. Observe the output on seven segment display for every clock pulse.

RESULT

The output of decade counter is visualized on seven segment display successfully.

PRECAUTIONS

1. Handle the ICs and electronic components carefully.
2. Check the ICs before starting of the experiment. Test the ICs using digital IC tester.

3. The connections should be neat and tight.

4. The LED has one leg long and other leg short. The long leg depicts positive end and short end depicts negative end.

5. Do not press the IC on breadboard until pins are aligned with pours properly.

6. Avoid short circuit and heating of ICs.

7. Ground should be common.

8. For data input +5V must be given for HIGH and 0V for LOW.

VIVA VOCE QUESTIONS

1. What is the significance of MOD counter?

2. Up to how many states, the MOD-N counter can count?

3. What is the significance of non-sequential counting?

4. What is the need of state table reduction?

5. Draw the state diagram of MOD-7 counter?

6. What is BCD counter? How many states it has?

7. Draw the timing diagram of any 4-bit counter?

Experiment 17

Digital to Analog Converter

OBJECTIVE

(A)   To design Weighted resistor D/A converter using Operational Amplifier.

(B)   To design R-2R Ladder D/A converter using Operational Amplifier.

APPARATUS

Breadboard, Power supply (+5V), LEDs, Resistor 1KΩ, 2KΩ, 4KΩ, 8KΩ, 10KΩ, 16KΩ, 20KΩ, Connecting wires.

IC REQUIRED

741 (OpAmp)

INTRODUCTION

The signals are used to carry information from one device to another. There are two types of signals for data communication, analog and digital. An analog signal is continuous wave that changes continuously over a period of time. The digital signals are electrical signals where data is converted into pattern of bits. It has discrete value at each sampling point. The other difference between analog and digital signals are form of representation. The analog signals are represented in form of sign waves, whereas the digital signals are represented in form of square waves. Human voice is an example of analog signal. CD, DVD, computers and digital devices, all are the examples of digital signals. The digital signals carry more information per second than the analog signals.

NEED OF CONVERSION

In real world application, most of the data information is characterised by analog signals. But the computers, microprocessor and machines understand digital languages i.e. data in the form of binary bit. So, it is required to convert the data from analog to digital form. Similarly, there are wide range of applications, where analog signals are required such as communication channels. So, it is required to convert digital signals into analog signals.

In real world, the signals are analog. These signals are continuous in nature. But, the processing of signals in digital environment is much easier, faster, efficient and less distorted by noise. So, conversion from analog to digital is needed. However, the output is again required in analog form. That's why digital to analog conversion is required.

In DAC, the digital signal processing is implemented before A/D conversion like encoding and analog signal processing is applied after the A/D conversion, for example, modulation. The fig. 17.1. shows the design flow of D/A conversion.

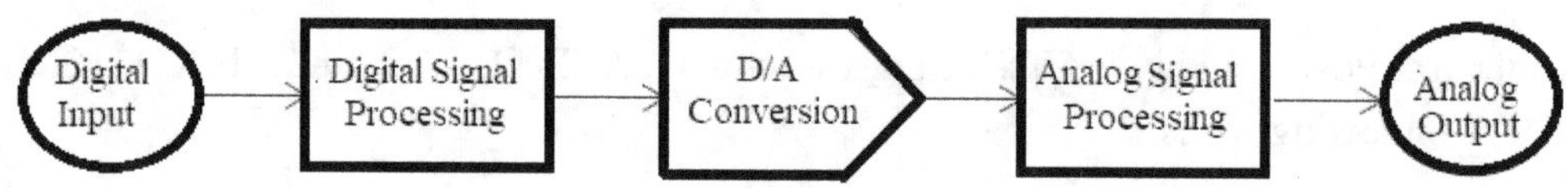

Fig. 17.1 D/A conversion design flow

Hence, as per user specifications and design environment, the ADC as well as DAC is required.

DIGITAL TO ANALOG CONVERTERS

A digital to analog converter is a device that converts digital signals (binary bits) into analog voltage or current output. According to the Nyquist-Shannon sampling theorem, any sampled data can be reconstructed perfectly with bandwidth and Nyquist criteria.

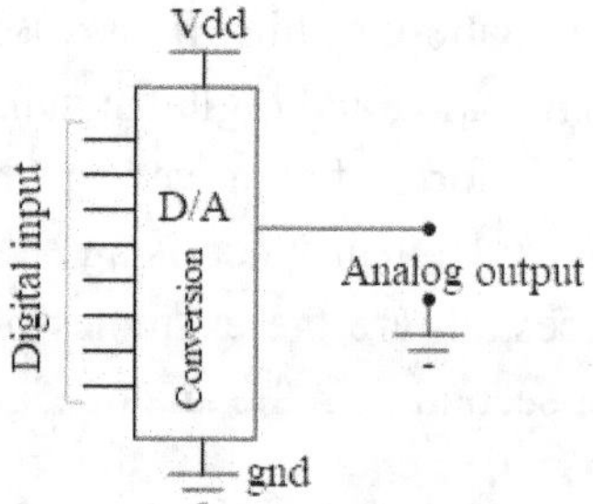

Fig. 17.2 Digital to analog conversion

The fig.17.2. shows the digital to analog conversion. There are basically two major types of digital to analog converters.

1. Weighted resistor D/A converter
2. R-2R ladder D/A converter

WEIGHTED RESISTOR D/A CONVERTER

The D/A converter constructed as weighted resistor uses a cascaded structure of R,2R,4R etc, DAC constructed as a binary-weighted DAC that uses a repeating cascaded structure of resistor values R and 2R. This improves the precision due to the relative ease of producing equal valued-matched resistors (or current sources).

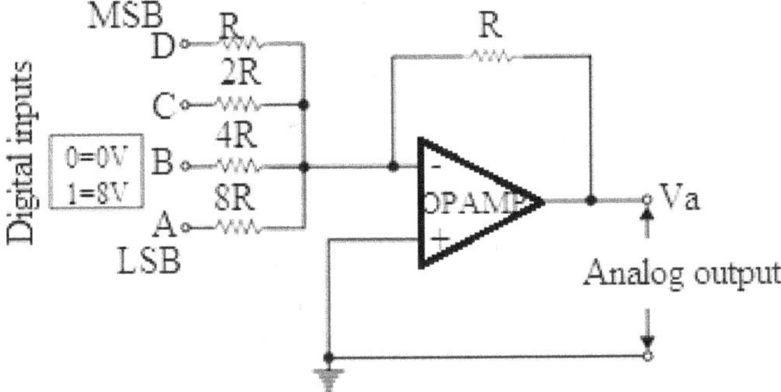

Fig. 17.3 Weighted resistor D/A converter

The fig. 17.3. shows weighted resistor D/A converter. Here an operational amplifier is used as summing amplifier. There are four resistors R, 2R, 4R and 8R at the input terminals of the operational amplifier with R as feedback resistor. The network of resistors at the input terminal of operational amplifier is called as variable resistor network. The four inputs of the circuit are D, C, B & A. Input D is at MSB and A is at LSB. Here we shall connect 8V DC voltage as logic–1 level. So, we shall assume that 0 = 0V and 1 = 8V. Now the working of the circuit is as follows.

Since the circuit is summing amplifier, its output is given by the following equation:

$$Va = -R(\frac{D}{R} + \frac{C}{2R} + \frac{B}{4R} + \frac{A}{8R}) \tag{17.1}$$

WORKING OF WEIGHTED RESISTOR D/A CONVERTER

When input DCBA = 0000, then putting these values in equation (17.1), we get:

$$Va = -R(\frac{0}{R} + \frac{0}{2R} + \frac{0}{4R} + \frac{0}{8R}) = 0 \text{ V} \tag{17.2}$$

When input DCBA = 0001, then putting these values in equation (17.1), we get:

$$Va = -R\left(\frac{0}{R} + \frac{0}{2R} + \frac{0}{4R} + \frac{8}{8R}\right) = -R\left(\frac{8}{8R}\right) = -1V \qquad (17.3)$$

When input DCBA = 0010, then putting these values in equation (17.1), we get:

$$Va = -R\left(\frac{0}{R} + \frac{0}{2R} + \frac{8}{4R} + \frac{0}{8R}\right) = -R\left(\frac{8}{4R}\right) = -2V \qquad (17.4)$$

When input DCBA = 0011, then putting these values in equation (17.1), we get:

$$Va = -R\left(\frac{0}{R} + \frac{0}{2R} + \frac{8}{4R} + \frac{8}{8R}\right) = -R\left(\frac{8}{4R} + \frac{8}{8R}\right) = -3V \qquad (17.5)$$

In this way, when digital input changes from 0000 to 1111 (in BCD style), output voltage (Va) changes proportionally. This is given in the conversion chart, as per table 17.1.

Table 17.1 Conversion table of weighted resistor D/A converter

Digital inputs				Analog output Voltage (Va)
D	C	B	A	
0	0	0	0	0V
0	0	0	1	−1V
0	0	1	0	−2V
0	0	1	1	−3V
0	1	0	0	−4V
0	1	0	1	−5V
0	1	1	0	−6V
0	1	1	1	−7V
1	0	0	0	−8V
1	0	0	1	−9V
1	0	1	0	−10V
1	0	1	1	−11V
1	1	0	0	−12V
1	1	0	1	−13V
1	1	1	0	−14V
1	1	1	1	−15V

There are some main disadvantages of the circuit, which are as following:

1. Each resistor in the circuit has different value.
2. Error in the value of each resistor adds up.
3. The value of resistor at MSB is the lowest. Hence, it draws more current.
4. Also, its heat & power dissipation is very high.
5. There is the problem of impedance matching due to different values of resistors.

R-2R LADDER D/A CONVERTER

It is modern type of resistor network. It has only two values of resistors the R and 2R. These values repeat throughout in the circuit. The operational amplifier is used at output for scaling the output voltage. The working of the circuit can be understood as follows. For simplicity, we ignore the operational amplifier in the above circuit (this is because its gain is unity). The fig. 17.4. shows R-2R ladder D/A converter.

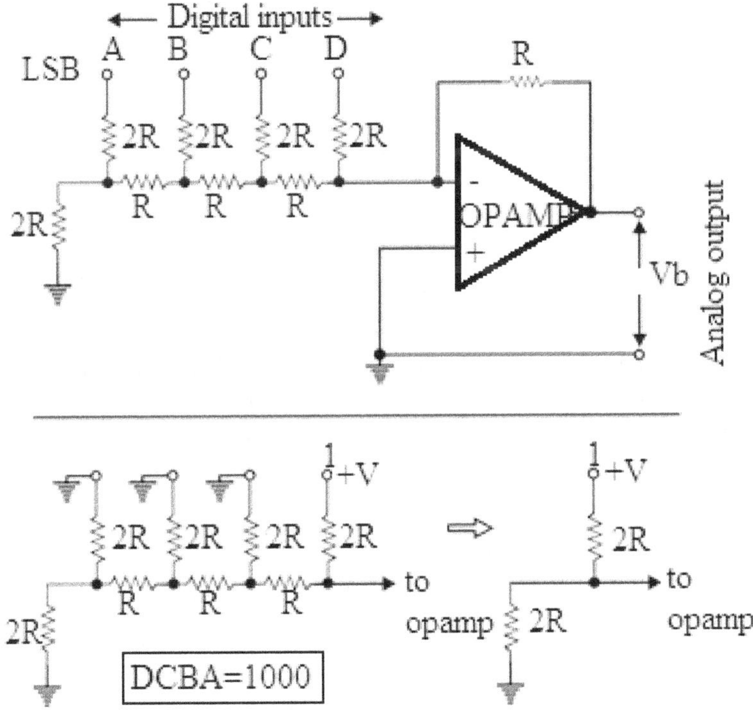

Fig. 17.4 R-2R ladder D/A converter

Now consider the circuit, without operational amplifier. Suppose the digital input is DCBA = 1000. Then the circuit is reduced to a small circuit.

Its output is given by:

$$Vb = \left(\frac{2R}{2R+2R}\right) \times (+V) = \frac{V}{2} \tag{17.6}$$

Now suppose digital input of the same circuit is changed to DCBA = 0100. Then the output voltage will be V/4, when DCBA = 0010, output voltage will be V/8, for DCBA = 0001, output voltage will be V/16 and so on. The general formula for the above circuit of R–2R ladder, including the operational amplifier also, will be:

$$Vb = -R\left(\frac{D}{2R} + \frac{C}{4R} + \frac{B}{8R} + \frac{A}{16R}\right) \tag{17.7}$$

Solving the equation,

Finally, we get:

$$Vb = -R\left(\frac{V}{2} + \frac{V}{4} + \frac{V}{8} + \frac{V}{16}\right) \tag{17.8}$$

With the help of this formula, we can calculate any combination of digital input into its equivalent analog voltage at the output terminals.

IC 741 (OPERATIONAL AMPLIFIER)

Operational amplifier is used as summing amplifier. An Operational Amplifier, or op-amp for short, is fundamentally a voltage amplifying device designed to be used with external feedback components such as resistors and capacitors between its output and input terminals. These feedback components determine the resulting function or "operation" of the amplifier and by virtue of the different feedback configurations whether resistive, capacitive or both, the amplifier can perform a variety of different operations, giving rise to its name of "Operational Amplifier".

An Operational Amplifier is basically a three-terminal device which consists of two high impedance inputs. One of the inputs is called the Inverting Input, marked with a negative or "minus" sign, (–). The other input is called the Non-inverting Input, marked with a positive or "plus" sign (+). The pin diagram of IC741 is shown as in fig. 17.5.

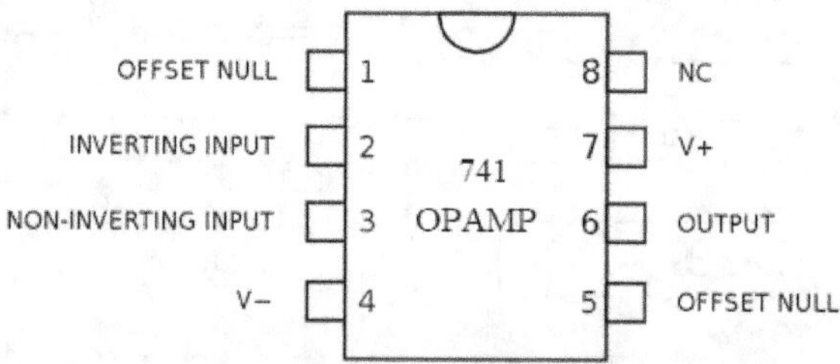

Fig. 17.5 Pin diagram of IC741

D/A CONVERTER ICS

1. The AD5755 is a complete, multichannel control IC that incorporates on-chip dynamic power control with four precision 16-bit programmable voltage, or 4 mA to 20 mA current, output DACs.
2. The AD5791 is the industry's first single chip DAC to feature true 1 ppm resolution and accuracy, providing 4× greater accuracy and 4× more resolution than competing converters.
3. The DAC8740H is Texas instrument industry's first modem to integrate HART, FF and PA protocols.
4. The DAC8775 is 16-bit quad-channel programmable current-output and voltage-output DAC, from TI.
5. The DAC70508 is Texas instrument industry's smallest high-performance DAC supporting 8 channels
6. The DAC0800 is 8-bit D/A converter. It has low power consumption and wide supply range.

EXPERIMENTAL PROCEDURE

1. Check all the ICs using IC tester. After inserting IC carefully in ZIP socket of the tester, the tester should show "PASS".
2. Carefully insert the ICs in breadboard.
3. To design 4-bit weighted resistor D/A converter, connect the circuit as shown in fig. 17.3.
4. Take reference voltage V_R as 5V.
5. Check analog output voltage V_a with help of multimeter.
6. Verify the output values observed with theoretical values.
7. To design R-2R Ladder D/A converter, connect the circuit as in fig. 17.4.
8. Take reference voltage V_{REF} as 5 volts.
9. Take analog voltage V_b with the help of a multimeter and verify the observation obtained with the theoretical values.

RESULT

(A)   The Weighted resistor D/A converter is designed using Operational Amplifier.

(B)   The R-2R Ladder D/A converter is designed using Operational Amplifier.

PRECAUTIONS

1. Handle the ICs and electronic components carefully.
2. Check the ICs before starting of the experiment. Test the ICs using digital IC tester.
3. The connections should be neat and tight.
4. The LED has one leg long and other leg short. The long leg depicts positive end and short end depicts negative end.
5. Do not press the IC on breadboard until pins are aligned with pours properly.
6. Avoid short circuit and heating of ICs.
7. Ground should be common.
8. For data input +5V must be given for HIGH and 0V for LOW.

VIVA VOCE QUESTIONS

1. What is the significance of converters?
2. What do you mean by 'offset null'?
3. What is resolution, accuracy and reliability?
4. What is Quantization and encoding?
5. Give the applications of D/A converter?
6. What is offset error and settling time?
7. Name any five digital to analog ICs?

Analog to Digital Converter

OBJECTIVE

To design Weighted resistor A/D converter using Operational Amplifier.

APPARATUS

Breadboard, Power supply (+5V), LEDs, Resistor 1KΩ, 220Ω, Connecting wires.

IC REQUIRED

74138 (Encoder), 741 (Operational Amplifier)

INTRODUCTION

The signals are used to carry information from one device to another. There are two types of signals for data communication, analog and digital. An analog signal is continuous wave that changes continuously over a period of time. The digital signals are electrical signals where data is converted into pattern of bits. It has discrete value at each sampling point. The other difference between analog and digital signals are form of representation. The analog signals are represented in form of sign waves, whereas the digital signals are represented in form of square waves. Human voice is an example of analog signal. CD, DVD, computers and digital devices, all are the examples of digital signals. The digital signals carry more information per second than the analog signals.

NEED OF CONVERSION

In real world application, most of the data information is characterised by analog signals. But the computers, microprocessor and machines understand digital languages i.e. data in the form of binary bit. So, it is required to convert the data from analog to digital form.

Similarly, there are wide range of applications, where analog signals are required such as communication channels. So, it is required to convert digital signals into analog signals.

In real world, the signals are analog. These signals are continuous in nature. But, the processing of signals in digital environment is much easier, faster, efficient and less distorted by noise. So, conversion from analog to digital is needed. However, the output is again required in analog form. That's why digital to analog conversion is required.

The fig. 18.1. shows the design for A/D conversion. The analog signal processing is implemented before A/D conversion like filtering and digital signal processing is applied after the A/D conversion, for example, error correction techniques.

Fig. 18.1 A/D conversion design flow

TYPES OF A/D CONVERTER

The A/D converter is the device that changes analog input to digital output. There are basically four types of A/D converters that are widely used in consumer electronics as well as industrial automation.

1. Flash type A/D converters

2. Counter type A/D converters

3. Successive approximation A/D converters

4. Dual slope A/D converters

Each type of converters has its own characteristics as well as application area along with some pros and cons.

2-BIT FLASH TYPE A/D CONVERTER

Flash type A/D converters are also known as parallel A/D converters. It is the fastest A/D converter. It is based on the principle of comparing analog input voltage with reference voltages. For n-bit conversion, 2^n-1 comparators are required.

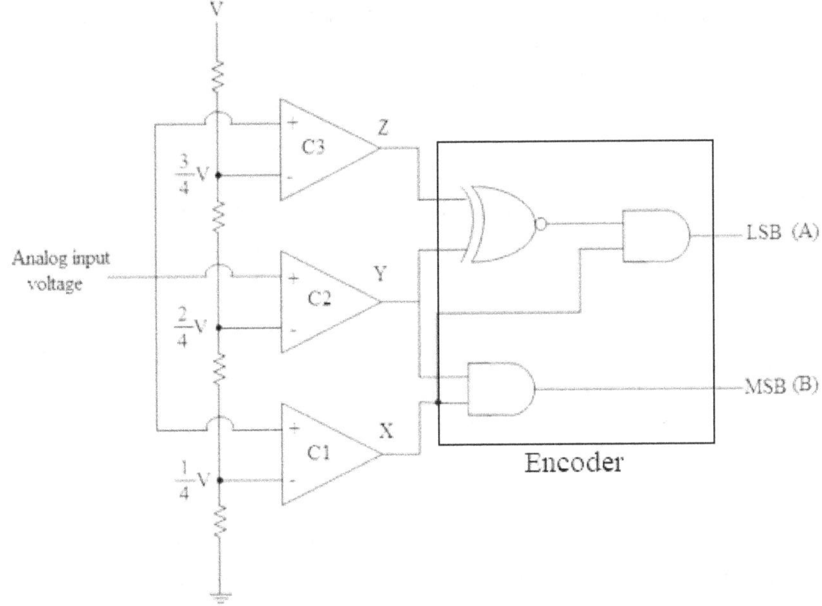

Fig. 18.2 2-bit Flash type A/D converter

To encode the comparator output to digital output, we need an encoder circuit.

The fig. 18.2. shows 2-bit flash type A/D converter. As n=2, so number of comparators will be 2^2-1=3. Here, operational amplifiers act as comparators. There are three comparators C1, C2 and C3. The inverting terminals of comparators are connected with reference voltages $\frac{V}{4}, \frac{2V}{4}, \frac{3V}{4}$ through resistive divider network and power supply V. The analog input signals are connected with non-inverting terminals of comparators. The table 18.1. specifies various possible analog conditions and corresponding comparator and digital outputs.

Table 18.1 2-bit flash type A/D converter

Analog conditions	Comparator outputs			Digital outputs	
	X	Y	Z	B	A
$0 \leq Vin \leq \frac{V}{4}$	0	0	0	0	0
$\frac{V}{4} \leq Vin \leq \frac{2V}{4}$	1	0	0	0	1
$\frac{2V}{4} \leq Vin \leq \frac{3V}{4}$	1	1	0	1	0
$\frac{3V}{4} \leq Vin \leq V$	1	1	1	1	1

When the voltage at inverting terminal is higher than the voltage at non-inverting terminal, it is at logic 0 i.e. negative saturation level. When the voltage at inverting terminal is lesser than the voltage at non-inverting terminal, it is at logic 1 i.e. positive saturation level.

Case (I): $0 \leq Vin \leq \dfrac{V}{4}$

As the Vin is more than 0 and less than $\dfrac{V}{4}$, so the output of all the comparators are 0 i.e. XYZ=000, corresponding digital output is BA=00, after implementing through logic gates.

Case (II): $\dfrac{V}{4} \leq Vin \leq \dfrac{2V}{4}$

As the Vin is more than $\dfrac{V}{4}$ and less than $\dfrac{2V}{4}$, so the output of all the comparators are 0 i.e. XYZ=100, corresponding digital output is BA=01, after implementing through logic gates.

Case (III): $\dfrac{2V}{4} \leq Vin \leq \dfrac{3V}{4}$

As the Vin is more than $\dfrac{2V}{4}$ and less than $\dfrac{3V}{4}$, so the output of all the comparators are 0 i.e. XYZ=110, corresponding digital output is BA=10, after implementing through logic gates.

Case (IV): $\dfrac{3V}{4} \leq Vin \leq V$

As the Vin is more than $\dfrac{3V}{4}$ and less than V, so the output of all the comparators are 0 i.e. XYZ=111, corresponding digital output is BA=11, after implementing through logic gates.

Thus, the analog input signal is converted into equivalent digital output.

IC 74148 (ENCODER)

We need to connect a Zener diode (5V) in reverse bias at the output of each IC74148, used as comparator, as in fig. 18.2.

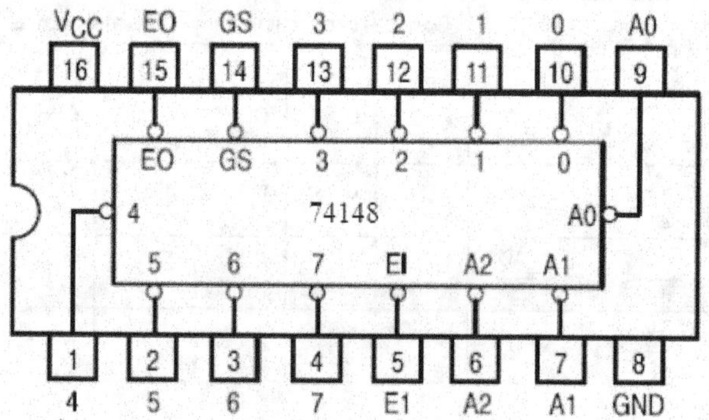

Fig. 18.3 Pin diagram of IC74148

CHARACTERISTICS OF FLASH TYPE A/D CONVERTER

1. As the conversion time is 100 ns. So, this is the fastest A/D converter and it is also known as parallel type A/D converter.
2. The construction is simple and easier to design.
3. It is limited to small number of bits, because the number of comparators will be increased. So, the operational speed will be low.
4. The power consumption is more.
5. It is well suited for high bandwidth applications.
6. This A./D converter does not require clock. So, the conversion time depends on the propagation delay and settling time of comparators.

APPLICATIONS OF A/D CONVERTERS

A/D converters are used for wide range of applications in consumer electronics and industrial automation. Few applications of A/D converters are listed as below:

1. It is used in digital voltmeters and cell phones.
2. It is used for data acquisition.
3. It is used in thermocouples.
4. It is used in digital oscilloscope.
5. It is used in microcontrollers.
6. It is used in voltage to frequency converters.

EXAMPLES OF A/D CONVERTER ICS

1. The ADC0808 is 8-bit microprocessor compatible, with 8-channel multiplexer A/D converters, from Texas Instruments.
2. The MCP3002 has a 10-bit analog to digital converter (ADC) with a simple to use SPI interface. It is manufactured by Microchip Technology.
3. The TMPM369FDFG contains two units of 12-bit sequential-conversion analog/digital converters (ADC) with 16 analog input channels, from Toshiba.
4. The ADC0804 is 8-bit analog to digital converter, which works at 0-5V analog input voltage. Its access time is $135\ ns$. It is manufactured by Texas Instruments.
5. The MCP3008-I/P is 10 bit, 8-channel, A/D converter from Microchip technology. Its operating voltage is 2.7-5.5V.
6. The ADS830E is 1- channel, 8-bit analog to digital converter, which works at 4.75-5.25V analog input voltage. Its sampling speed is $60\ Msps$. It is manufactured by Texas Instruments.

EXPERIMENTAL PROCEDURE

1. Check all the ICs using IC tester. After inserting IC carefully in ZIP socket of the tester, the tester should show "PASS".
2. Carefully insert the ICs in breadboard.
3. To design 2-bit flash type A/D converter, take reference voltage V_R as 5V
4. Take 1KΩresistor for each of resistors in the voltage divider.
5. Make connections of the circuit as shown in fig. 18.2.
6. Make the observations as per table 18.1.

RESULT

The 2-bit flash type A/D converter is designed and verified successfully.

PRECAUTIONS

1. Handle the ICs and electronic components carefully.
2. Check the ICs before starting of the experiment. Test the ICs using digital IC tester.
3. The connections should be neat and tight.
4. The LED has one leg long and other leg short. The long leg depicts positive end and short end depicts negative end.
5. Do not press the IC on breadboard until pins are aligned with pours properly.
6. Avoid short circuit and heating of ICs.
7. Ground should be common.
8. For data input +5V must be given for HIGH and 0V for LOW.

VIVA VOCE QUESTIONS

1. What is the significance of ADC and DAC?
2. In 8-bit converter, if sampling rate is 1ms and full range scale is 5V. Then find
   (a) number of possible states   (b) resolution
3. Explain the Nyquist-Shannon theorem of sampling?
4. What are the two main factors on which accuracy of ADC rely?
5. What do you mean by aliasing? When it occurs?
6. What do you mean by term repeatability in A/D converters?
7. How do you measure resolution in ADC or DAC?

List of Innovative Projects based on Digital Electronics

1. To generate Hamming code, detect and rectify errors using flipflops and basic gates.
2. To Design a digital calculator which can implement addition and multiplication functions, and display output in 7-segment display unit.
3. To Design a digital calculator which can implement subtraction and division functions, and display output in 7-segment display unit.
4. To Design a universal counter which can perform different shift operations using multiplexer.
5. To Design a up and downfading lights (different coloured LEDs) with specified delays using flipflops/counters.
6. To Design Universal shift register using IC74194.
7. To Design 3 bit Binary to Gray code converter using logic gates.
8. To Design BCD counter using logic gates.
9. To Design 8 to 3 priority encoder using logic gates.
10. To Design a Line Following Robot using basic Logic gates.
11. To Design digital fan speed regulator using ICs and gates.
12. To Design digital Bank token number display using ICs and gates.
13. To Design digital fan regulator using ICs and gates.
14. To Design digital counter for an auditorium using ICs and logic gates.
15. To Design 3-bit counter and display the output in the 7-segment display.